Education Division 2022

AA002976

Held at the 2022 AIChE Annual Meeting

Phoenix, Arizona, USA
13-18 November 2022

ISBN: 978-1-7138-7895-7

Printed from e-media with permission by:

Curran Associates, Inc.
57 Morehouse Lane
Red Hook, NY 12571

Some format issues inherent in the e-media version may also appear in this print version.

Copyright© (2022) by AIChE
All rights reserved.

Printed with permission by Curran Associates, Inc. (2023)

For permission requests, please contact AIChE
at the address below.

AIChE
120 Wall Street, FL 23
New York, NY 10005-4020

Phone: (800) 242-4363
Fax: (203) 775-5177

www.aiche.org

Additional copies of this publication are available from:

Curran Associates, Inc.
57 Morehouse Lane
Red Hook, NY 12571 USA
Phone: 845-758-0400
Fax: 845-758-2633
Email: curran@proceedings.com
Web: www.proceedings.com

TABLE OF CONTENTS

EDUCATION DIVISION AWARDS PLENARY (INVITED TALKS)

FREE FORUM ON ENGINEERING EDUCATION: FIRST YEAR AND SOPHOMORE YEAR

FREE FORUM ON ENGINEERING EDUCATION: JUNIOR AND SENIOR YEARS

IN HONOR OF THE 2021 RECIPIENT OF THE WARREN K. LEWIS AWARD - NICHOLAS PEPPAS - PART I (INVITED TALKS)

NSF WORKSHOP I: HIGHLIGHTS FROM CBET

NSF WORKSHOP II: PROPOSAL WRITING AND DISCUSSIONS WITH PROGRAM MANAGERS

POSTER SESSION: CHEMICAL ENGINEERING EDUCATION

SURVEY RESULTS AND BEST PRACTICES (INVITED TALKS)

NEW FACULTY FORUM (INVITED TALKS)

ASSESSMENT AND EVALUATION OF STUDENT LEARNING AND PERFORMANCE

COMPUTING AND DATA SCIENCE IN CHE EDUCATION

SUPPORTING STUDENT DEVELOPMENT OF PROFESSIONAL SKILLS AND RESPONSIBILITIES

SAFETY IN CHE: HOW CHE DEPARTMENTS ARE SATISFYING THE AICHE PROGRAM CRITERIA

SOCIETAL ISSUES IN THE CHE CLASSROOM AND CURRICULUM

MODERNIZATION AND INTEGRATION OF THE CHE CURRICULUM

BROWN-BAG LUNCH-AND-LEARN: TEACHING TIPS

WORKSHOP: ROUNDTABLE DISCUSSIONS ON CURRENT TOPICS IN CHE EDUCATION

CHEMICAL ENGINEERING LABORATORY EXPERIMENTS AND INSTRUCTION

546a Innovating Real World Global Challenges in Chemical Engineering Education

Courtney Pfluger

It is crucial to have engineering students understand the complexities of solving real world global challenges as they will be the ones innovating solutions for the future. This means as educators, we need to provide students with learning experiences that get them to think about not just the engineering concepts to a solution but also the environmental, social, cultural, and economic ramifications of their proposed solutions. Translating these types of learning outcomes to the classroom is important as educators to give the students authentic experiences and projects to facilitate this cross-cultural learning and informing students of the global context to large societal issues. To achieve this, there is also a need to teach and foster inclusivity as they work in teams of people from diverse backgrounds in the future. This talk will discuss the ways I have implemented the concepts of inclusion, innovation, and global context thinking into Chemical Engineering courses I teach and the global programs I have developed.

546b Professional Service: A Rewarding and Cautionary Tale

Anthony Butterfield

The helping profession aspects of engineering can bring a great sense of purpose and joy to the chemical engineering workplace. An engineer can often discover and implement small improvements, and then see them directly benefit a great many lives, when implemented on a large scale. Professional service gives a similarly motivating and addictive aspect to our occupations. In this talk I will discuss lessons learned through a history of working with our engineering community to find those small but impactful improvements through professional service. I will discuss strategies that I have found effective for getting involved in and creating new ground in professional service activities. Service, though, can go wrong--mistakes have been made--and this talk will end with reflections on how to avoid some tempting professional service pitfalls that I wish I had avoided.

546c Envisioning the Future of Chemical Engineering Education

Cheryl A. Bodnar

As the chemical industry continues to evolve and change to meet the needs of today's society, it is essential that we prepare chemical engineering students for the diverse set of challenges that they will encounter. Our students need to recognize opportunities, apply their technical knowledge, build professional skills such as communication, leadership, and teamwork, and use creative thinking to solve complex problems. One pedagogical approach that can help motivate our students while developing the aforementioned skillsets is game-based learning. Games are beneficial in that they can immerse learners within diverse contexts while pushing them to work to the edge of their capabilities. Games also have the additional benefit of being able to provide safe spaces for students to experience failure without encountering the real world consequences resulting from choices made. This seminar will discuss the use of both physical and digital based games for the education of chemical engineering students. Research into the effectiveness of the game implementations will be shared along with best practices for use of these game-based strategies in the classroom.

237a Teaching Chemical Engineering with Coffee: Lessons Learned with 1,800 Students per Year

William Ristenpart, Tonya L. Kuhl

We describe a class developed at U.C. Davis titled 'The Design of Coffee,' which serves as a nonmathematical introduction to chemical engineering as illustrated by the process of roasting and brewing coffee. Hands-on coffee experiments demonstrate key engineering principles, including material balances, chemical kinetics, mass transfer, conservation of energy, and fluid mechanics.

The experiments lead to an engineering design competition where students strive to make the best tasting coffee using the least amount of energy - a classic engineering optimization problem, but one that is both fun and tasty. 'The Design of Coffee' started as a freshmen seminar in 2013, and it has exploded in popularity: this academic year it is serving more than 1,800 students as the largest and most popular elective course at U.C. Davis.

In this talk we focus on the class pedagogy as applied to hands-on activities, with an emphasis on how coffee serves as an engaging and exciting topic for teaching large numbers of students about engineering design in an approachable, hands-on manner.

237b Using an Aeropress Coffee Brewer to Teach Fluid Mechanics

Kristen E. Fawole, Kristen Fawole, Tonya L. Kuhl, William Ristenpart

Fluid mechanics is a cornerstone of chemical engineering, but courses in fluid mechanics typically require students to have prior knowledge of advanced mathematical and physical concepts. The steep prerequisites complicate efforts to create entry-level, introductory fluid mechanics pedagogy for engineering or non-engineering students. Here, we describe a laboratory module called "Pressure Driven Flow through Coffee Grounds", which is offered as a part of a general education, no-prerequisite class titled "The Design of Coffee: An Introduction to Chemical Engineering", a course which regularly enrolls over 500 students each academic quarter at the University of California Davis. In this laboratory session, students perform hands-on experiments with an "AeroPress" brewer that allows students to simultaneously measure the manually applied pressure difference and the resulting flow rate of brewed coffee. After the experiment, students are expected to have learned how to: (1) relate the concepts of pressure and flow rate using Darcy's Law, (2) assess how the flow rate affects the strength and quantity of the brewed coffee, and (3) clearly communicate technical data via graphs. We measured gains in achievement of these first two learning objectives using quizzes administered before and after the lab session, and we assessed the third learning objective by comparisons of submitted graphs versus a standardized rubric. We present statistical analyses of achievement in each category, and we further discuss how the AeroPress module can be incorporated into existing first-year classes tailored to different cohorts of students.

237d Attracting Sophomores to Careers in Process Control and Automation through a Winter Term Workshop and Targeted Internships

Jason Berberich, Douglas Coffin, Gary Rudemiller, Patrick Dixon, Keith Hohn

The fourth industrial revolution which includes artificial intelligence, automation, control, and big data are transforming all aspects of industry. An outcome of this technological transformation is a shift in the skills required for new and practicing engineers. Today many manufacturing companies are challenged to find engineers equipped with the skills needed to work in a manufacturing plant with modern controls and automation technology. In response to these emerging needs we have developed a program with targeted industrial internships in the areas of process control and automation which provide an avenue to introduce students to this field earlier in their education. The internship program includes an intensive training workshop that students complete before undertaking a sponsored internship.

The workshop, titled *Systems Automation Springboard to Internships* (SASI), is a three-week program taking place during the winter term with the goal to present students with the full spectrum of automation, from field instrumentation to enterprise control systems. Participating students were sponsored by companies, who paid the workshop fee, and offered the students an internship during the summer. The pilot workshop was offered in January 2021 and the second cohort just finished the workshop this January 2022. The target audience of the SASI program is second-year engineering students so we can introduce students to process control early in their education. The workshop is taught by a mix of instructors from academia and industry. The instructors from industry provide job specific examples and perspectives on skills-gap knowledge in control and automation. The workshop also provides an intensive hands-on experience in controls through lab exercises such as practice in preparing and interpreting PFDs and Pandamp;IDs, performing regression and fitting data to models, tuning controllers, and programming PLCs.

The value of the program is to improve student's awareness of the control and automation field to enhance internship experience, and provide employers with the opportunity to attract a skill-specific engineering workforce. Feedback from the sponsor companies and students give high marks for the workshop. The success of the program is highlighted by the doubling of participating students offered internships in the two years that the program has been offered.

237e Mastery Learning in the Mass Balances Class

Payton Kamer, Glenn Lipscomb

The first year of the chemical engineering curriculum is critical to student success. The principles of mass and energy balances (MEB) taught then are the basis for problem solving in virtually all subsequent classes.

The desire to have students demonstrate a minimum level of mastery has led to the introduction of mastery learning in MEB. Mastery learning is based on the premise that all students can achieve a desired level of understanding if given sufficient time and resources. Resources include classroom instruction material as well as individual and group tutoring with mentors, graduate teaching assistants, and instructors.

For MEB, the mass balances content was divided into individual learning concepts, e.g., unit conversion and solving mass balances for a recycle problem with reaction. A series of problems were developed for each concept and deployed as quizzes in the Blackboard course management system. Random values are assigned to specific variables so a single correct answer does not exist.

Students must obtain the correct answer to receive credit for a quiz. No partial credit is given. Since the quizzes constitute 72% of the grade, successful completion of the quizzes is required to obtain a passing grade. If a student does not pass a quiz on their first attempt, they may retake it an unlimited number of times. Between attempts, students are encouraged to seek help through a review of instructional material or help from others.

Student responses have been favorable. They appreciate having unlimited time to work problems. They also appreciate the opportunity to review concepts and correct errors if they do not obtain the correct answer.

Blackboard provides summary data on the number of attempts and time per attempt for each quiz. This information has been used to develop two metrics of student learning that the instructor can use to improve student performance:

1. Conceptual Understanding
2. Solution Effort

Differences in the percentage of students who complete the quiz on their first attempt (as opposed to multiple attempts) indicate how well students understand the concepts. If most students complete the quiz on their first attempt, current instructional approaches are effective in preparing students. Otherwise, modifications may be beneficial.

Differences in the time required to obtain the correct solution indicate the effort required to solve the problem. The total reflects both the time required to transform the problem statement into a mathematical expression and the time to solve the equations. Providing an anticipated solution time for students may help students prepare mentally and allot sufficient time to complete the quiz thereby fostering self-motivation.

Results for both metrics are presented and use of the data to improve instruction discussed.

237f Relating "Mass and Energy Balances" Concepts to Everyday Life and Developing Professional Skills

Cody Mischel, Betul Bilgin, James Pellegrino, Lewis Wedgewood, Vikas Berry

As part of our NSF RIEF project *"Developing Professional Identity: Integrating Academic and Workplace Competencies within Engineering Programs",* students are asked to present concepts taught in the "Mass and Energy Balances" course (CHE 210) and relate them to everyday life. Students formed teams of five and were asked to choose a concept and everyday problem from either a proposed list or of their own creation. Upon approval of their topic by the course teaching assistant, they prepared a 5-minute PowerPoint presentation and created a video. During the Fall 2021 semester, students only prepared a PowerPoint presentation and presented it in the class to their peers and two industry mentors who were present. Industry mentors evaluated their presentations and gave them feedback. In the Spring 2022 semester, students presented to the rest of the class and industry mentors and additionally, they created videos. A standardized rubric was used for assessment which includes these elements: understanding of principles that govern daily phenomena, design of the presentation and video, clear explanation of the concept, and delivery of the material. The project team collected evaluation data from mentors. Moreover, qualitative and quantitative data will be collected from students, course instructors, and industry mentors to measure the effectiveness and impact of this implementation on student learning. Once videos are finalized and reviewed by the instructor, a YouTube channel will be created, which can help cultivate engineering identity while also providing a sense of ownership for the UIC CHE community. Our goal is to develop intrapersonal and interpersonal skills in a fun environment while also fostering professional development by direct engagement with the industry. By executing this task, we aim to create an *in-situ* engineering environment resulting in students' direct engagement with industry supporting the formation of professional identity. These tasks will help students to identify themselves as engineers as early as sophomore year. Implementation of these tasks in *CHE 210* can be a driver of change and transformation of departmental education culture in the future. Moreover, videos posted on YouTube publicly will serve to promote UIC engineering programs. If this intervention is shown to be efficient and impactful, it can be easily scaled to other courses.

237g Teaching and Learning Strategy of Aspenone during COVID-19 Pandemic to Chemical Engineering Sophomores

Sushobhan Pradhan, Sundararajan Madihally

The COVID-19 (SARS-CoV-2) pandemic has drastically and negatively influenced students learning since mid-Spring 2020 and therefore, it gave rise to a sudden shift to the online education delivery system from the conventional in-person teaching method. Even though students moved to several virtual platforms during the pandemic, it was a serious challenge to provide hands-on training to teach and train various laboratory skills. In a dry laboratory setting, the students are often facilitated with many modern and complicated devices to conduct various computational or applied mathematical analysis. The process simulation courses have become an extensive and essential tool in chemical engineering and is becoming an important part of the chemical engineering curriculum. In addition, the process simulation and computational expertise facilitate an association of research and teaching. Nevertheless, during the pandemic period, the virtual learning mode often prevents exposure to such hands-on training facilities for teaching such process simulation courses. In this article, we have discussed an online delivery mechanism to teach two AspenONE software i.e., Aspen Plus and Aspen Hysys, for CHE 2033: Introduction to Chemical Process Engineering course, which is offered to sophomores in the School of Chemical Engineering at Oklahoma State University (OSU), Stillwater, Oklahoma, USA. In addition, we discussed the experience of teaching blended online delivery mechanisms (synchronous and asynchronous) of process simulators. We believe that the experiences learned during the COVID-19 pandemic will benefit in developing novel rules, regulations, learning platforms and potential solutions for upcoming unavoidable circumstances that adversely affect the established instructional mechanisms.

237h Imperial Chemical Engineering Wiki: The Development, Lessons Learned and Future Growth of Student-Led Wikis in Engineering Education

Luc Paoli, Thomas Nok Hin Cheng, Pierre Walker, David -, Marsha Maraj, Jerry Y. Y. Heng

Wikis enable community-driven dissemination of knowledge, characterized by the freedom for any user to edit and create content. In recent times, wikis have become a prevalent learning tool in higher educational programs where these are often staff-led and integrated into specific classes. However, a review of the literature shows that there is limited evidence of student autonomy in the development of wikis.

In this talk we introduce the Imperial ChemEng Wiki, chemengwiki.com, which is a student-led initiative from within the Department of Chemical Engineering at Imperial College London. This had the initial aim of formalizing a structure for the effective sharing of student-scribed notes and to provide students additional opportunities for increased collaboration and peer scaffolding. The ChemEng Wiki was formally launched in 2020 and has run for the past three academic years with contributions from over 30 undergraduate student developers. During this period, Google Universal Analytics was used to collect usage data which highlight a widespread use, recording approximately 55,000 sessions from over 50 countries.

Surveys were run periodically, and on average 90% of the students who responded found the Wiki to be a valuable means of retrieving information, consolidating concepts, and preparing for their examinations. Although not an immediate response to the COVID-19 pandemic, the ChemEng Wiki has also shown the potential to mitigate isolated student experiences and enhance collaborative learning among students.

Feedback from the student developers was crucial towards the iterative improvement of the Wiki. We aim to empower others to foster similar initiatives by reporting both the lessons learned during each stage of the development and the framework and resources that have enabled this initiative to succeed.

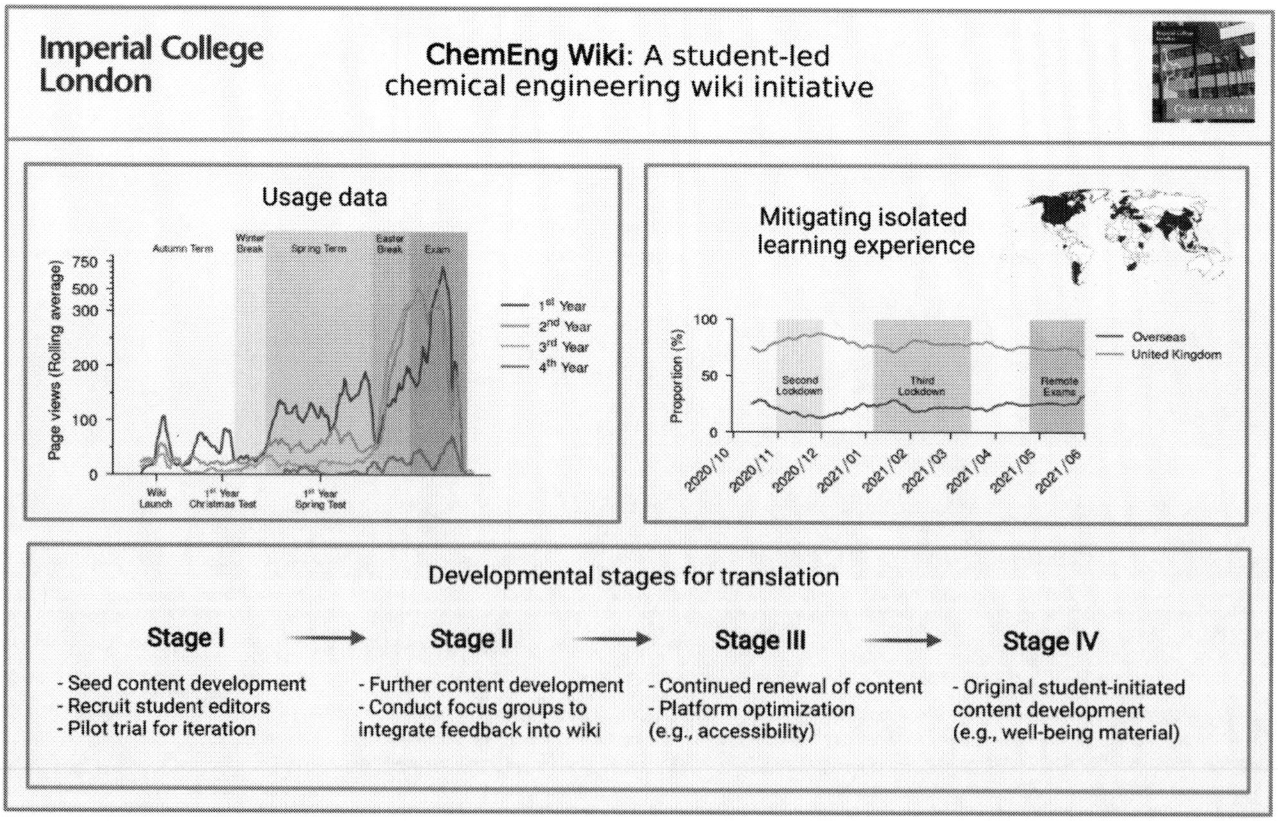

48a Teaching Fugacity through Comics and Assessing the Impact on Student Confidence and Understanding

Lucas Landherr, Francisco Hung, Rebecca L. Carrier

Comics have been used as learning tools in engineering education for several years with positive feedback from students and educators. The comics have been developed by integrating written lectures and instruction with visual depictions to provide new curricula to help students better grasp more abstract and complex topics, effectively providing students with 'visual notes' that allow them to revisit the lectures where the concepts were originally taught and helping them to form deeper connections with the subject matter. In a number of general studies, integrating the comics into instruction or simply providing them as additional tools have been determined to increase the students' interest, confidence, and understanding of the concepts addressed.

One particular 10-page comic, developed in collaboration with a professional artist, focused on the subject of fugacity. This comic has since been used by a number of educational institutions, as well as international federal agencies and even as supporting evidence in a legal case to help explain the concept to a jury. It has clearly been received well given its broad adoption; however, the impact of the comic has never been truly assessed in a formal educational setting. In order to evaluate the comic, students in two thermodynamics sections were provided the comic at the end of their fugacity instruction. Students were provided a series of ConcepTest questions focused on important features of fugacity in gases, liquids, and mixtures, in order to best measure their understanding. Following each question, students were asked to assess their confidence in the answer they provided. Students then read through the comic, before answering the ConcepTest questions again and providing another assessment of their confidence. Differences in their answers and self-assessment were then evaluated to determine the effectiveness of the comic.

This paper discusses the evaluation of the comic with recommendations on the best means to integrate a comic learning tool into thermodynamics courses for greatest positive impact.

Northeastern

48b The Impact of an Optimization Mindset Focused on Long-Term Student Success: A Junior-Level Chemical Engineering Course Case Study

Kristen M. Wilding, Hayley Ford, Brad Bundy

As optimization experts, engineering faculty are uniquely capable of applying an optimization mindset to their courses. However, education is one of the most difficult optimization challenges. Each student has a different background and learning preferences and each faculty has their own pedagogical strengths and weaknesses. Herein, we report results from a recent study published in Chemical Engineering Education[1] where an optimization mindset is applied to a Junior Level Chemical Engineering Course. The objective of the optimization experiment is to focus on college activities that are correlated with long-term student success in careers and in life (citizenship, leadership, service) within the predominant constraints of engineering education (time, financial resources, and student/faculty motivation and abilities). The results of an IRB-approved multi-year study are reported herein. Overall, applying an optimization mindset to the design course experiences was well received by students and led to increased student engagement.

Reference:

1) Wilding KM, *Ford H*, Bundy BC. 2021. Applying an Optimization Mindset to Engineering Education: Junior Level Course Project Case Study. *Chemical Engineering Education*. 55(4):228-234. https://doi.org/10.18260/2-1-370.660-114933.

48c Replacing Mimicry with Mastery: Homework Wrappers and Effort-Based Grading

Carl Lund

In a problem-solving course, students often use a mimicry approach to solve homework problems. Their approach is to find a solved problem that appears to be similar and to mimic the steps taken in that solution to solve the assigned problem. The mimicry approach is maladaptive, inefficient (in terms of learning) and often ineffective. In this study, the solution to each assigned homework problem was provided to the students as a part of the homework assignment. Students were instructed to attempt to solve the problem as if they were taking an exam. They were told that the purpose of the homework was for them to gauge their mastery for solving that type of problem, and that their submission would be graded exclusively on the basis of their effort to master the solution process and not on whether they obtained the correct answer. They were told to refer to the provided solution only if they 'got stuck' while solving the problem, and then to use it after they had completed their solution to determine whether they had made any mistakes. Each assignment included a homework wrapper question. This question asked them to reflect upon their mastery for solving this type of problem, to identify any misunderstandings that had resulted in mistakes, and to determine what they should do to remove misunderstandings and further develop their mastery for solving this type of problem. The student responses to the wrapper questions, together with observations of their help-seeking activities, suggested reduced reliance on pure mimicry during problem solving. In cases where significant mimicry was observed, there still were indications that their achievement goals may have shifted to include understanding how to solve the problem and not just getting the correct answer.

The context of this study was a required, upper-division kinetics and reaction engineering course. Between 2005 and 2015, the course was revised in a number of ways. It was converted from didactic lecture to a flipped classroom format. Explicit instruction was added in the areas of identifying the problem type and enumerating a general approach for solving each problem type. Instead of jumping directly to what the instructor did to solve each problem, the instructor explained how they knew what to do. 'Mastery coaching,' that is messaging from the instructor stressing that the goal is to know how to solve a problem, not just getting the answer, was introduced throughout class meetings. This also included the instructor noting that getting stuck on a problem was good because it showed a point where the student's understanding was not yet complete. Scaffolding was added to the in-class problem-solving activities each time a new reactor type or problem type was encountered. Finally effort-based grading of homework and the inclusion of homework wrapper questions in assignments was introduced.

As noted, a flipped classroom format was used to teach the course. Prior to class, students were expected to read text and/or watch videos that presented theory, derived and simplified equations and demonstrated how to use that theory and equations to solve a variety of different reaction engineering problem types. The problem types included parameter estimation (fitting models to reactor data), reactor response, process optimization and system design. During class, the pre-class information was briefly reviewed and students were afforded an opportunity to ask questions. Most of the time in class the students were engaged in problem-solving activities. A single homework problem (including the solution and a wrapper question) was due before the start of the next class. Some homework problems included a 'thought question' that was intended to make the students interpret or qualitatively explain the answer to a problem. The homework submissions were awarded one point for an attempt to set up the equations need to solve the problem, one point for attempting to perform the calculations and obtain the answer, one point for answering the wrapper question and one point for answering the thought question; the correctness of the answers were not checked. At the end of the course, these homework scores were scaled to represent 20% of the students' final course grade. Importantly, 70%

of the students' final course grade was based upon evaluations where they were graded on correctly setting up the necessary equations and performing the calculations to obtain a correct answer, but the students had always been afforded ample opportunity to practice prior to these evaluations.

When wrappers were first introduced in the course, they were more focused upon whether the student had correctly identified the problem type and, if they made a mistake, where in the general approach for solving that problem type they had made the mistake. In addition, the solution to the problem was made available and the wrapper question was assigned after the students submitted their solutions. There were some encouraging results (the recurrence rate of mistakes that were identified using wrappers was 4.6% compared to 29.3% for those that were not identified). However, three observations prompted the change to include the solution and wrapper with the assignment. First, 31% of the wrapper responses were deemed to be insincere (the students did not take them seriously). Second, the wrapper submission rate was only 61%. Finally, there was no evidence to suggest that the use of mimicry had diminished.

The fall 2021 offering of the course (it is only offered in the fall) was the first time the problem, wrapper and solution were all provided at the time of assignment. The wording of the wrappers was also changed to emphasize mastery over identifying solution process mistakes. As an example, here is a wrapper used with a parameter estimation problem: 'The second learning outcome for [course number] is the ability to estimate values for parameters in a postulated rate expression and assess its suitability using kinetics data from an ideal stirred tank or plug-flow reactor. This is the first [course number] kinetics data analysis problem. After you have completed your solution, take a few minutes to reflect upon your current ability to satisfy this learning outcome. Consider whether you would easily recognize problems of this type on an exam, whether you know how to proceed without consulting solved examples of this type of problem, your overall confidence in your abilities for this type of problem, what you might do to increase your understanding and confidence, etc. Then write a short paragraph summarizing your reflections.'

The wrapper submission rate increased slightly to 69%, but far fewer insincere responses were submitted. More importantly, the responses indicated students were focusing on problem-solving mastery and not only on getting the correct answer. In particular, the responses frequently assessed whether or not the respondent felt they could solve a similar problem under exam conditions (e. g. without access to a solved problem). Responses to successive wrappers on problems of the same type typically evidenced increasing mastery. Similarly, students often came to office hours to discuss their approach to solving a problem even though they had been provided with the solution.

At this point, the wording of the wrappers requires some minor revision, but they appear to be eliciting the desired student response. One focus in the next offering of the course will be on using the wrappers to initiate a dialog between the students and the instructor. The other focus will be measuring the students' achievement goal orientation and how it changes over the course of the semester. The present study suggests that students are moving toward a mastery orientation and away from a performance orientation. At this point, the combined use of effort-based grading and assignments that include both the problem solution and an appropriate wrapper question appears to be a promising approach to reducing student use of mimicry when solving homework problems and increasing their focus on the development of their problem-solving mastery.

48d A Descriptive Study of a New Course on the History of Thermodynamics

Rachel Morrish, Eliza Buhrer

A new course on the history of thermodynamics titled "*Thermo-Human Dynamics - Heat, Energy and Time*" was developed and taught in the Spring of 2020 and Spring of 2022. The class was co-instructed by Prof. Eliza Buhrer, a historian, and Prof. Rachel Morrish, a chemical engineer. This presentation provides a descriptive study of the course design as well as student perspectives and outcomes. The class itself examines how discovery of the 1st and 2nd Laws of thermodynamics not only transformed scientists' understanding of the universe, but also had a significant impact on nineteenth and twentieth century art, literature, religion, philosophy, and social theory. The course was run as a weekly, 3 hour discussion seminar interspersed with interactive exercises, student presentations, and writing workshops. Students were led through a series of historical and cultural readings, bringing to light often overlooked connections between science and the arts. Assignments included weekly writing exercises, an oral biography of a thermodynamicist, and a creative project artistically portraying a thermodynamics concept. We will discuss the advantages and challenges of developing a cohesive course bridging two disciplines, and how studying thermodynamics through a historical lens can deepen students' understanding of its fundamental principles.

48e Enhancing Collaborative and Inclusive Skills Development to Foster Effective Chemical Engineering Capstone Design Teams

Courtney Pfluger, Anastasia Hauser, Sindia M. Rivera-Jimenez

Capstone design is one of the quintessential courses in an undergraduate student's curriculum that requires effective teamwork skills to be successful. Effective teams foster innovation, empathy, safety, and belongingness while clearly communicating goals and distributing effort in order to solve a problem. These team attributes are learned skills that employers specifically look for when hiring engineers in the workforce so it is critical that effective teamwork skills are taught in the classroom, as emphasized in the 2020 ABET criteria. This presentation will discuss the application of inclusive and collaborative team pedagogies at three different institutions. These strategies include but are not limited to team charters/community agreements, project management, defined roles and responsibilities set by the teams, activities on effective and inclusive discussions, team peer assessments, and meaningful feedback. The effectiveness of these pedagogical strategies is evaluated using the Team Learning Beliefs and Behaviors Model, adapted into a questionnaire. In this model, team effectiveness is shaped by the individual's beliefs about the team's interpersonal context, leading to their willingness to engage in learning behavior. The five attributes of effective and inclusive team behaviors measured by the questionnaire are Interdependence, Social Cohesion, Group Potency, Psychological Safety, and Task Cohesion. The questionnaire was administered in all three universities before and after the capstone course. The results found that the students thought interdependence (leadership, delegating responsibilities, communication, and clear team roles as essential to effective teamwork), psychological safety (feeling safe and open to address team issues and project ideas), and task cohesion (setting goals and performing tasks to achieve a final project) were the most important attributes of effective teams. These results may indicate that instructors are effective at preparing students to be on task and to provide inclusive discussions within teams. However, it may also indicate that pedagogical methods to enhance team social cohesion and group potency with more targeted group team building activities could benefit team effectiveness.

48f An Approach to Addressing "Bio" Design within Capstone Design

Troy Vogel, Mark J. McCready

One of the more onerous courses in the chemical engineering curricula is capstone design. This course often has a significant role in identifying attainment levels of student outcomes for ABET accreditation. Often the 'catch-all,' this course is viewed as the 'last opportunity to catch a problem going out the door.' To add to this challenge, students enter, typically in their final year and sometimes with a job in hand, with more and more starts to their careers outside of traditional chemical process industries. With varying interest and commitment levels, some simply want to pass and graduate, others will earn the A however they can. One observation that the faculty of the chemical engineering program at Notre Dame has made is that students have an increased interest in the biological applications of the fundamentals of chemical engineering. This outlines an approach to teaching a capstone design course that allows for biological designs alongside traditional chemical processes. The agency afforded to students in choosing even bio vs non-bio can provide a stronger connection to the course content.

At Notre Dame, capstone design is a one semester course taken in a student's final graduating semester. This is the only named course taken in this semester and all others are electives of some variety. The course covers equipment design, flowsheeting, economic evaluation, and process safety with an emphasis on building review of fundamental core topics whenever practicable. The course is taught by two instructors, one with primary responsibility for the lecture component and one with responsibility for the design project.

Within team formation, students are able to elect if they want to work on a "bio" or a "non-bio" project. In the past two years, approximately 50% of the students have elected a bio project, 30% non-bio, and the remaining 20% indicated no preference. Teams are made using CATME with a very strong bias toward maximizing full team availability, aligning schedules to promote ease of out of class teamwork.

This talk outlines the approach used to address both bio and non-bio material within the lecture component of the course so that both bio and non-bio topics are benefited from the skills developed during the lecture. The work further outlines the types of bio projects that work well and share those we've found to be more problematic. The approach complies with the program accreditation procedure.

48g Working with Industry in Rowan's Engineering Clinic Program

Jacob Martin, Erik Dunn, Emily Rooney, Andrew Sikora, Kirti Yenkie, C. Stewart Slater, Robert Hesketh

The Rowan University engineering clinic program focuses on exposing undergraduate students to graduate level research with industrial applications. Many of these clinics work directly with industry to ensure the research and development they are focused on is relevant to the real-world and is providing value. This is a case study of our clinic project which focused on the lubricant manufacturing industry.

Lubricant manufacturing and processing facilities produce a variety of lube oils. These lube oils are often processed in batch on the same line. In order to remove the residual oil a process of pigging, draining, and then flushing is conducted. The flushing oil is normally the next oil being processed in the line. The result of this flushing process is that it results in two co-mingled oils that no longer meet the product specifications. This co-mingled oil is then downgraded to a lower value oil and is often sold for much less than the original oils, leading to losses for the manufacturer. The focus of this project is to optimize this process and minimize the losses from producing downgraded oil.

Through Rowan's clinic program, our team worked closely with a prominent industrial lube oil producer. An abundance of industrial experience was gained by interacting with industry engineers and scientists through regular meetings over the course of the semester. In addition, we also had several visits to the plant in order to understand and model the industrial process. In order to understand this process, we observed the flushing operation several times and developed a process flow diagram of the process which was refined and improved over several meetings. We also were given data on the flushing operation and even helped to collect some of the samples at the plant. Collaborating with an industrial partner gave us the opportunity to gain an understanding of the real-world applications of our learning alongside an early introduction to chemical engineering outside of academia.

One of the unique features of our clinic education is the ability to learn about an industry that is not normally covered in our classes. So to start on this project we had to search the literature to find out about the flushing problem in this industry and to find potential solution strategies. We generated a number of possible solutions and then discussed these with our industry partners. In these discussions with industry we were exposed to considerations that are important to the company that professors didn't put at high importance. We also received training from the industry in a Lube 101 course that they presented to us. In this presentation we learned the petroleum lube oil terminology and classifications. These classifications introduced the concept of "families" of oil which represents the various sets of desired properties that exist for different applications. We were more directly exposed to daily operations through plant tours with safety training that helped us understand the industry standards for employee safety. Additionally, these trainings and tours have allowed us to develop an understanding of industrial culture through interactions with operators, industry liaisons, and operations managers. This has shown the dynamics at play within the chemical industry giving us a heads up over those students who have not had this experience.

The application of a learned skill can be one of the most difficult lessons to impart on a student. The work we performed in this engineering clinic taught us how to apply what we have learned throughout our time as students. We have used fluid dynamics to mathematically model the forces at play during the mixing of oil. The mathematical modelling of the situation was the basis for the scale down calculations that designed the size of our pilot plant under construction. All this work was done in Excel spreadsheets. Excel is a powerful data computation tool which proved essential to the work performed in this clinic. Developing a greater understanding of the program and how to build spreadsheets for real world applications gives us a great advantage moving into industry.

Good communication is one of the most important traits needed in a good engineer. The ability to communicate results and ideas with a wide audience is an important skill. This project has given us the opportunity to communicate and give many presentations for various audiences including professors, technical industry contacts, and managerial industry contacts. In normal coursework we have mainly given technical presentations as our professors and fellow students are all from a technical background. However, in many cases members of an audience will not come from a technical background. Presentations need to be tailored to fit a specific audience, while still conveying the same results and ideas. This project has given us experience presenting to an audience of all backgrounds (chemists, engineers, managers, technicians) and has really helped develop our communication skills.

Our team has investigated many creative solutions of reducing oil holdup in piping systems such as air blowing, coated pipelines, and pigging. The challenge is how to test these solutions to determine their potential impact. In order to determine that, reliable methods were needed to determine oil purity. The primary method used in industry to determine the kinematic viscosity is the ASTM D445. This process requires that a sample be taken and then sent up to an analysis lab and an automated measurement is made. Then the result is communicated to the operator who either flushes for more time if the sample failed and if the sample passed then begins to fill the bottles or drums. Since this is a time consuming process we decided to investigate alternative methods to determine if we were done flushing. We investigated alternative ways to make this measurement without stopping the process. One method that has shown some process is a light absorbance measurement and another method is to make an online viscosity measurement.

One of the biggest benefits of our clinic project is it allows us to work on a real-world problem. In most classes our time is spent solving problems from a textbook that always have a clearly defined solution. This project allows us to really use our creativity to create and test solutions the problem. Not only does it enable us to be creative, but it allows us to apply the knowledge we have gained from our coursework. It give's us a chance to apply our knowledge and previous education in a way that prepares us for future jobs.

The project has also given us insight into how companies approach problems and how they search for solutions. For example, this project has a big focus on reducing financial losses by solving this issue. This has really changed how we as a team have approached the problem and given us more to focus on. The real-world work and insight into industry gives us great experience and a real advantage after graduation.

In this project we learned that we needed to know more information that we could by looking at the data provided by industry. So we designed and developed simple experiments has helped further the team's knowledge of research-based learning. For example, how much oil sticks to the wall of a pipe? Can oil be trapped in a filter housing? How long does it take for the flush oil to replace the residual oil? Along with many more questions. Working closely with an industrial partner has given the team excellent real-world experience, and a better understanding of the lube oil processing industry. Using modeling and computer software to aid data analysis helped further the team's understanding of useful skills and made connections from classroom to real world learning. Frequent technical presentations advanced the team's communication and soft skills. Creating these connections from industry and classroom knowledge has strengthened the team as engineers and helped prepare them for future endeavors.

48h Using Mediawiki As an Online Collaborative Tool to Foster Learning in the Transport Phenomena Sequence at the University of South Florida

Ryan Toomey

Chemepedia.org, a website powered by MediaWiki (a free and open-source wiki software platform), was developed as a collective learning tool for students enrolled in the Transport Phenomena sequence at the University of South Florida. It was developed so that students can participate in active learning through updating and contributing to the website.

The Transport Phenomena sequence is two courses taken in the Junior year. The sequence is largely broken down between a microscopic treatment of the transport phenomena (first class) and a macroscopic treatment of the transport phenomena (second course).

The Chemepedia.org website is organized into 26 learning modules covering both courses. Each module encompassing the following: 1) Overview; 2) Learning Objectives; 3) Learning Outcomes; and 4) Practice Exercises that correspond to the learning outcomes.

Each semester, students are given assignments involving either editing or adding content. In particular, students are either given assignments to develop and post questions that address learning outcomes in the learning modules or to edit and add to previously posted content. Over the past 3 years, the site has grown to over 300 pages as the result of efforts by students to present material in a manner that facilitates their learning.

This talk will address experiences, observations, and outcomes of these efforts.

238a Teaching Better Biomedical Engineering: Sixty Years of Chemical Engineering Contributions That Shaped up Bioengineering

Nicholas Peppas

The Principles of Chemical Engineering became extremely important in the young field of biomedical engineering 65 years ago. The 1960s and 1970s were a period of wild scientific expansion on the chemical engineering principles to the world of biomedical and pharmaceutical sciences in a way that nobody could have expected or predicted. Through the pioneering contributions of the late Ed Merrill of MIT, Mitchel Litt of U Penn, Alan Michaels of MIT and Stanford, Ken Bischoff, then of the Universities of Texas and Maryland, Bob Dedrick of NIH, Kumar Chandrasekaran of ALZA Corporation, Clark Colton of MIT, John Quinn and John Anderson of Illinois, Don Paul of the University of Texas, Robert Langer of MIT, Ken Himmelstein of the University of Kansas, Michael Sefton of the University of Toronto, Ping Lee of Ciba-Geigy and now University of Toronto, and the writer, all chemical engineers, the principles were set for the quantitative development and rational design of the new generation of biomedical devices, advances artificial organs, simple and complex medical units, tissue engineered products and drug delivery devices that are common now. Chemical engineering provided the solid background that led to this design and optimization and became the early work of what Adam Heller and others call now a '*system-reponsive medical device*'. A key problem of biology and medicine this century has been to reduce the problems of disease to problems of molecular science and fundamental engineering problems. Many of the associated methodological advances in biomedical sciences are the result of earlier investments in the basic sciences. Thus, breakthroughs in molecular science and ChE principles have opened a floodgate of new opportunities for curing disease. In this work (i) we summarize the impact of chemical engineering in BME; (ii) we analyze major societal changes and how they impacted the development of BME; (iii) we point out the significant impact of ChE principles in solving advances medical problems and formulated 21st century bioengineering; and (iv) we point out the early leadership of chemical/biomedical engineers in the shaping of diversity, equity and inclusivity in the field.

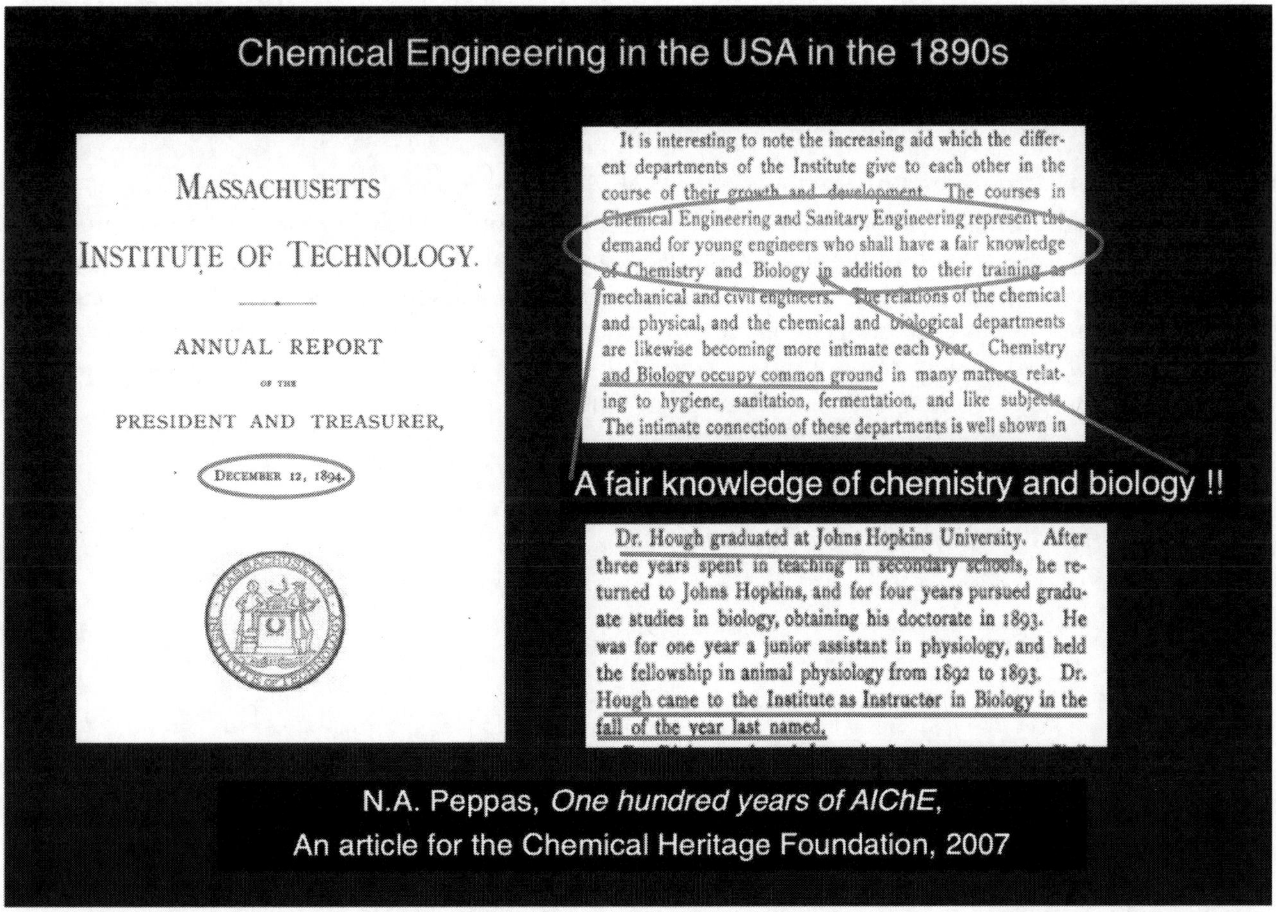

Chemical Engineering in the USA in the 1890s
MASSACHUSETTS INSTITUTE OF TECHNOLOGY.
ANNUAL REPORT OF THE PRESIDENT AND TREASURER,
DECEMBER 12, 1894.
It is interesting to note the increasing aid which the different departments of the Institute give to each other in the course of their growth and development. The courses in Chemical Engineering and Sanitary Engineering represent the demand for young engineers who shall have a fair knowledge of Chemistry and Biology in addition to their training as mechanical and civil engineers. The relations of the chemical and physical, and the chemical and biological departments are likewise becoming more intimate each year. Chemistry and Biology occupy common ground in many matters relating to hygiene, sanitation, fermentation, and like subjects. The intimate connection of these departments is well shown in
A fair knowledge of chemistry and biology !!
Dr. Hough graduated at Johns Hopkins University. After three years spent in teaching in secondary schools, he returned to Johns Hopkins, and for four years pursued graduate studies in biology, obtaining his doctorate in 1893. He was for one year a junior assistant in physiology, and held the fellowship in animal physiology from 1892 to 1893. Dr. Hough came to the Institute as Instructor in Biology in the fall of the year last named.
N.A. Peppas, One hundred years of AIChE,
An article for the Chemical Heritage Foundation, 2007

238b Regenerative Engineering Education and Democratization of Science: The Uconn-M1 Program

Cato Laurencin

UConn REM-M1 is designed to create leaders in engineering research and education from underrepresented groups. The program is built upon the vision of Dr. Laurencin and combines three innovative activities. 1. Mentoring and Training of Individuals to become leaders capable ultimately of National Science Foundation Funding through focused and sustained research experiences. 2. The promotion of racial/ethnic diversity in leadership ranks in engineering. 3. Educating individuals on the Convergence field of Regenerative Engineering with the goal of creating from the ground up, a field with leaders who are diverse, and an environment that is inclusive.

238c Sowing the Seeds of Diversity in Engineering Via Mentoring

Jennifer Curtis

The lack of women and underrepresented minorities in engineering is a problem that no single solution can address. Not only are there the pipeline issues, but the challenges these groups face in navigating their careers need to be addressed. This presentation will share some activities/practices regarding **mentoring** at every career stage that are instrumental to sowing the seeds of diversity.

238d Perspectives and Suggestions on How to Advance Underrepresented Minority (URM) Faculty at Minority Serving Institutions

Carlos Rinaldi-Ramos

The speaker will share perspectives and suggestions on how faculty at predominantly white and research-intensive institutions can contribute to diversity and inclusion by providing mentorship and helping the advancement of underrepresented minority (URM) chemical engineering faculty at minority serving institutions (MSI), beyond the usual circles formed by formal mentor-mentee relationships and academic family trees. These perspectives are based on the speaker's own experiences as a URM faculty and as faculty at a Hispanic Serving Institution, as well as conversations with other URM faculty colleagues throughout the years.

109a Division Director Update on the Chemical, Bioengineering, Environmental, and Transport Systems Division of NSF: Programs and Opportunities

Jeanne VanBriesen

The mission of the **Chemical, Bioengineering, Environmental and Transport Systems (CBET)** Division of NSF's Directorate for Engineering is to support innovative research and education in the fields of chemical engineering, biotechnology, bioengineering, and environmental engineering, and in areas that involve the transformation and/or transport of matter and energy by chemical, thermal, or mechanical means. CBET research and education investments contribute significantly to the knowledge base and to the development of a 21st century workforce for major components of the U.S. economy, including chemicals, pharmaceuticals, medical devices, forest products, metals, petroleum, natural gas, food, textiles, energy utilities, alternative energy sources, microelectronics, and other sectors. Support for environmental research encompasses pollution prevention and remediation as well as life-cycle analysis. This talk will highlight Foundation, Directorate, and Division level opportunities that may be of interest to the AIChE community of researchers.

109b Briefing of the Chemical Process Systems Cluster of the CBET Division of the National Science Foundation

Ray Adomaitis

The Chemical Process Systems (CPS) cluster of programs in CBET supports research seeking innovations in catalytic design, reaction engineering, multiphase chemical separations processes, molecular thermodynamics, process control and design, and sustainable energy conversion. The four programs within this cluster also participate in a variety of division-, directorate-, and foundation-wide funding initiatives. This talk will describe the programs in the cluster, highlight notable accomplishments achieved in supported projects, and discuss funding opportunities relevant to the CPS research community.

109c Update on Programs in the Transport Phenomena Cluster of the CBET Division of NSF

Shahab Shojaei-Zadeh

The Transport Phenomena Cluster at CBET supports fundamental research on the transport of mass, momentum, energy, and species. There are four core programs in this cluster: Fluid Dynamics, Particulate and Multiphase Processes, Combustion and Fire systems, and Thermal Transport Processes. Research proposals supported by these core programs span across several disciplines including Chemical, Mechanical, Aerospace, Petroleum, and Environmental Engineering.

109d Update on Programs of the Engineering Biology and Health Cluster of the NSF CBET Division

Steven Peretti

The Engineering Biology and Health (EBH) cluster of programs in CBET supports a diverse research portfolio, ranging from nano-scale molecular design to macroscale assistive device development. Programs within this cluster also participate in a variety of division-, directorate-, and foundation-wide funding initiatives. This talk will describe the programs in the cluster, highlight notable accomplishments achieved in supported projects, and discuss funding opportunities relevant to the EBH research community.

109e Update on Programs of the Environmental Engineering and Sustainability Cluster of the CBET Division of NSF

Mamadou Diallo

The Environmental Engineering and Sustainability (EES) Cluster of the NSF/ENG/CBET Division is organized around three core programs: 1) Environmental Engineering, 2) Environmental Sustainability, and 3) Nanoscale Interactions. This presentation will give an overview of the core programs of the CBET EES Cluster. We will also highlight crosscutting and convergence research opportunities within the EES cluster with a focus on recent division-, directorate-, and foundation-wide initiatives relevant to the EES community.

187a Proposal Writing Strategies for Fundamental Research Proposals for NSF

Triantafillos Mountziaris, Dimitrios Papavassiliou

This workshop, conducted by former program officers of NSF's Engineering Directorate, will discuss strategies for putting together fundamental research proposals, typical of the National Science Foundation. The workshop will discuss the importance of motivation of the problem, research approach, and describing the intellectual merit and broader impacts of the project. The workshop will be followed with informal networking with NSF program officers at the meeting.

366a Gamification of Learning: Designing a Safety Assessment Card Game

Seng Hoe (billy) Hue, Elton M. Dias

'Edutainment', or educational entertainment has been around since the 1970's, from the simplicity of solving a crossword puzzle on the back of a cereal box to the implementation of Virtual Reality in the subject of complex fluid mechanics. In a pedagogical environment, how do educators ensure that the information they are passing on is not lost in the midst of lengthy texts typically found in any academic subjects?

Although safety and safety assessments are key components of any Chemical Engineering degree, these topics are often overlooked by students, who typically find them tedious, uninteresting and intangible. We have designed an educational card game, '0 Days Since', to teach students about the most common safety assessments used by chemical engineers in different industries. Historic industrial accidents are a core part of the game, highlighting the importance of safety assessments while adding a realistic dimension to the game. The objective of the game is to assemble a team of safety assessment experts and tackle industrial crises, ultimately preventing the safety sign to be reset to '0 days since the last incident' (as the title suggests). While the topic is transversal to chemical engineering, this project directly supplements many undergraduate modules, serving as a refresher on safety assessments. This game was designed to be asymmetrical, balanced, competitive and enjoyable while being educational with no prior knowledge in engineering safety required. During a game, players are also exposed to other interesting non-engineering notions such as Tort law or a SEC spot-check, giving the players a glimpse of real-world events and activities that go beyond the typical core engineering concepts taught in classes.

From designing artwork to setting the rules, undergraduate and postgraduate students were involved along the way to ensure that the depths of the topic are right, while keeping the intrinsic motivations of the game. The artwork was individually crafted by two undergraduate students, taking in their own flair and creativity in interpreting each concept and card. This card game was funded by UCL Changemakers, which is part of UCL 2034's long-term strategy for student-staff partnership, bridging the gap between teaching and learning. Additionally, we will explore more into the benefits of a student-staff partnership, provide key learnings on game design with a breakdown for each step, and tips and tricks for maintaining the sensitive balance between an educational and an entertaining card game.

0
DAYS
SINCE
OIL RIG BLOWOUT
Crisis
Updated Regulations
Play this card after any player (including you) rolls the dice
+2/-2
Hazard &
Operability
Each time you roll for a Legal Dispute, +2 to your roll.
Risk Assessment
Secretary
CEO

366b The Oral Exam – an Alternative Approach to Assessment

J. Patrick Abulencia

A traditional paper exam is one of the most artificial ways to assess an individual's ability because there is only a single opportunity (i.e. you can only write one answer, which is judged right or wrong), information is unidirectional, and it is not typically performed outside of academia. A better approach would include a dialogue between student and professor where incorrect responses can be immediately corrected, and students given an opportunity to resume the correct logic. To address this, an oral exam format was used as an assessment tool in both the Material and Energy Balances and Thermo II courses. Briefly, students were asked to generate a slide deck and present the most important concepts for a particular section of the course. The instructor would periodically interject during the presentation to ask a student to provide a deeper explanation, if it was not initially done so. The first utilization of this method in Thermo II solely relied on the presentation, but the recent implementation in Material and Energy Balances incorporated a problem recitation by the student. Reactions to this methodology will be presented, as well as example content and the rubric used for evaluation.

366c Data-Driven Analysis of Learning Behavior within a Student-Led Chemical Engineering Wiki

Thomas Nok Hin Cheng, Luc Paoli, Samuel Chun-Hei Lam, David -, Pierre Walker, Jerry Y. Y. Heng, Marsha Maraj

The ChemEng Wiki is a student-led initiative aimed at improving collaborative learning and fostering technical expertise within the chemical engineering cohort at Imperial College London. The primary functions of the Wiki are to enable students to share their interpretations of taught material with other students and both develop and disseminate new content effectively.

The Wiki has been operational for the past two academic years, following a pilot in early 2020. During this period we have had the opportunity to evaluate the impact of the platform using pseudo-anonymized usage data through Google Universal Analytics. Analysis of the temporal dataset obtained via applying machine learning techniques, such as dimensionality reduction, has uncovered substantial heterogeneity in the way the Wiki was used to facilitate learning throughout the cohort. This has allowed us to correlate differences in learning behavior. In particular we highlight clusters of students who access material across subject years (cross-reading) and who access material during semester-time versus non-term time (pre-reading). These clusters help to better understand how students access material, how they use to this build their engineering competence and how the format and framework of the Wiki can be improved.

Such analysis provides the basis of developing an integrated open-source data collection and analysis pipeline, which represents a novel way of approaching web usage analytics within higher education. This further demonstrates the untapped potential a data-driven approach holds in aiding the development of an authentic appreciation of student learning, and provides evidentiary support towards the use of wikis in higher education pedagogy.

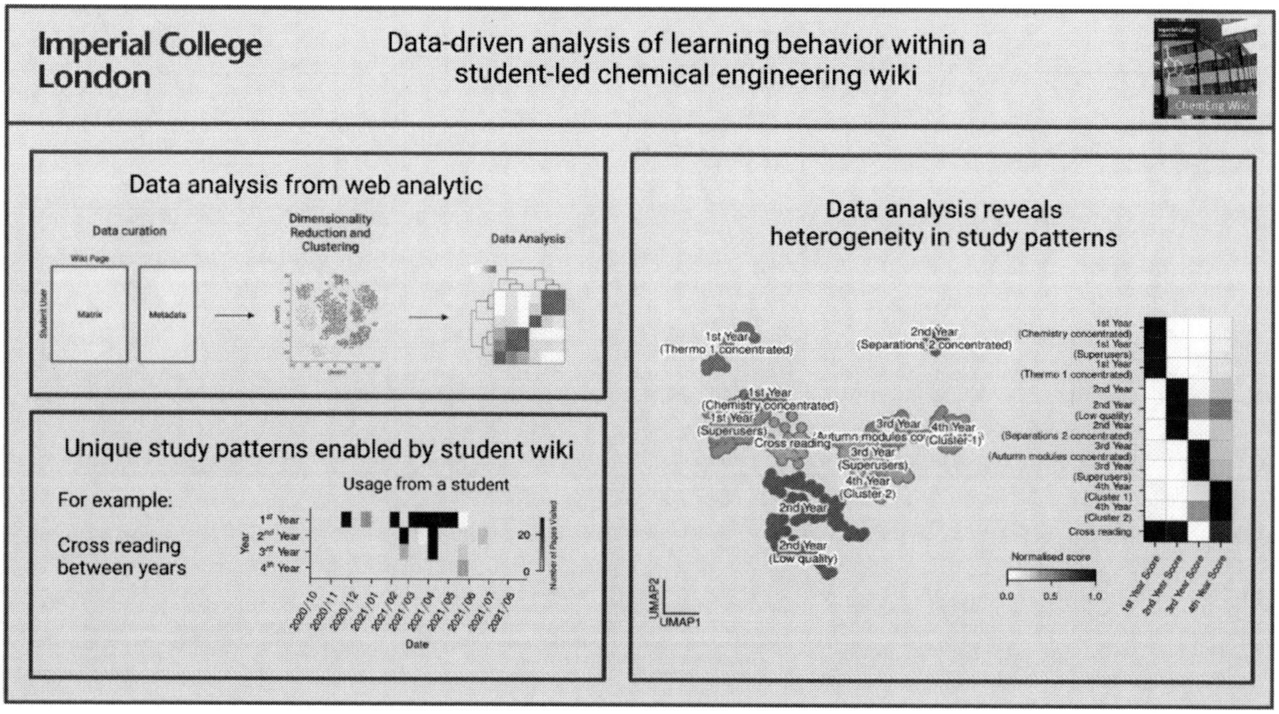

366d Making Unit Operations an Immersive Experience through Augmented Reality

Jacob Crislip, Esai Lopez, Cameron Armstrong, Andrew R Teixeira

A chemical engineering student's knowledge of theory, experimental design, and real-world processes is tested in the Unit Operations laboratory. Consequently, instructors' resources are stretched increasingly thin in pursuit of delivering high-quality content. In-lab time is too often spent distributing personalized instruction regarding the experiment at hand. Thus, inefficient use of class time restricts the breadth and quantity of data students are able to collect thereby diminishing their opportunity for learning. More efficient delivery of course content will enhance familiarization, operation, and safety of lab equipment facilitating critica l thinking as students navigate unfamiliar experimental conditions. This work demonstrates the evolution of several technological tools, including augmented reality, used to effectively train students on key aspect of unit operations before and during the lab sessions.

366e Translating DNA Origami Nanotechnology to Middle School, High School, and Undergraduate Laboratories

Anjelica Kucinic, Peter Beshay, Patrick Halley, Carlos E. Castro, Michael Hudoba

DNA nanotechnology is an emerging field enabling researchers to create nanoscale constructs and tools that have potential to advance many fields. Specifically, DNA origami is a rapidly evolving nanotechnology that allows researchers to create geometrically precise and complex nanostructures for applications in fields such as molecular sensing, robotics, and nanomedicine. The interdisciplinary nature of the development and applications of DNA origami can provide many benefits to students exposed to basic knowledge and methods of the nanotechnology. However, the cost of materials and equipment and complexity of design and characterization of nanostructures limits DNA origami experiments mainly to research focused institutions in graduate level laboratories with prior expertise and well-equipped laboratories. This work seeks to overcome crucial barriers to translating DNA origami methods to educational and non-graduate laboratories such as middle-school and high school science classes and primarily undergraduate institutions.

The most significant barriers are related to resources and time required for DNA origami fabrication. The equipment needed ranges from possibly available (e.g. gel electrophoresis), to unlikely available (e.g. thermocyclers), or impractical (e.g. electron microscopy) for ready access in educational settings. The barrier of time comes from all stages of the process including: 1) Design – which may take up to weeks for novice users to learn the design process or even learn the design software; 2) Fabrication – DNA origami fabrication may take several hours to prepare and up to days for thermal annealing; 3) Analysis – analysis methods like gel electrophoresis typically requires 2-4 hours to setup and run and microscopy imaging is likely impractical for educational settings. In addition to these barriers the complexity of the design and fabrication process are challenging to educational translation. Here, we present a streamlined protocol for fabrication and analysis of static and dynamic DNA origami nanostructures that can be conducted within a 2-hour laboratory course using low-cost equipment that contains readily available materials in educational laboratories and science classrooms.

The first educational experiment module focuses on a static DNA origami nanorod structure that was previously developed for drug delivery applications. The second educational experiment module focuses on a nanomechanical DNA origami structure initially designed to provide a framework and model for deformable, or compliant, nanostructures. These modules provide a protocol for students to analyze structures via gel electrophoresis using classroom-ready gel equipment. The modules include learning tools related to DNA nanotechnology such as charge screening, mechanical deformations, conformational dynamics and free energy landscapes, nanoscale stimulus response, polymerization, and algorithmic design and assembly. Additionally, the nanomechanical DNA origami educational modules provide a framework to illustrate how simple beam bending mechanical models can be used to predict nanostructure properties, as a class extension. The modules provide topics that relate to a variety of subjects including engineering, chemistry, physics, biology, materials science, medicine, and computer science, and hence could be worked into a range of curricula.

Over the long-term, introducing a broad range of students to DNA origami would also have the potential to advance the field due to increased interest and involvement by young students, who may then pursue education, research, or career paths related to DNA nanotechnology. This work focuses on eliminating major barriers that make DNA origami fabrication and experiments challenging to perform in educational laboratories and classroom environments. Here we overcome these barriers to enable the hands-on introduction and use of DNA origami technology in instructional labs and classrooms. Furthermore, introducing students to DNA nanotechnology and related fields can also have the potential to increase interest and future involvement by young students.

366f Sense of Belonging within Chemical Engineering and Its Impacts upon Retention

Duncan Mullins

This is a project examining chemical engineering identity and its relation to student retention and persistence. Chemical engineering is a strenuous discipline, filled with challenging academic coursework, difficult problems to focus upon post graduation within professional careers, and rigorous internal standards in chemical engineering classrooms and societies. Furthermore, within engineering educational settings often there is a celebration of suffering, a culture of embracing the struggles of the program from the students. This is not only prevalent in chemical engineering, it is accentuated due to the amount of chemistry courses necessitated in most degree programs. It is known that retention rates within engineering programs in the USA struggle, disproportionately affecting minoritized groups, underrepresented groups, and students belonging to a lower socioeconomic status. These include women, LGBTQIA students, and students of non-white identitiy.

It is known that identity and belonging can be major intangible factors influencing retention and resilience within academic or professional contexts. However, little work has been done on students' sense of belonging within chemical engineering or their internal identity as an engineer and how it relates to their resilience or persistence within the program. When students struggle with the rigor of the programs, often they are at risk or placed upon academic probation. Academic probation is a primarily punitive measure, often rocking the identity of students placed upon probation or at risk of probation.

Consequently, we want to investigate how identity and belonging relate to persistence and retention of chemical engineering students who are academically “at risk”, as defined by being on or at risk of being placed upon academic probation, or “struggling”, by self identification. We define persistence and retention as continuing through the semester, specifically continuing through the program to the third year, or 5th semester consecutively.

We conducted interviews with chemical engineering students who fit our criteria of being on academic probation or self-identifying as struggling within the program, seeking to understand their identity and whether there was a sense of belonging within their academic program, and how that related to their willingness to persist/work to get off of probation and achieve within the remainder of their program. The students we interviewed were at least 2nd semester, and we examined qualitatively the data of these interviews as well as the metrics of how many of our participants continued in the program the following year/semester. This data will be analyzed through thematic coding(1), examining how students consider their identity and their feelings of belonging within the academic program via the metric developed within Godwin’s (2016) facets of identity within engineering (2).

We expect the implications of this study to be potential ramifications on retention, specifically of those currently underrepresented groups noted earlier, within chemical engineering, and can be a justification for programs to improve sense of belonging and foster engineering identity within students.

1. V. Braun and V. Clarke, “Using thematic analysis in psychology,” Qualitative Research in Psychology, vol. 3, no. 2, pp. 77–101, Jan. 2006, doi: 10.1191/1478088706qp063oa.
2. A. Godwin, “The Development of a Measure of Engineering Identity,” in 2016 ASEE Annual Conference andamp; Exposition Proceedings, New Orleans, Louisiana, Jun. 2016, p. 26122.

   doi: 10.18260/p.26122.

366g Graduate Involvement in Vertically Integrated Projects for Biomaterials Education Research

Alec Svenson, Jennifer Weiser

Hands-on educational experiences during the first few years of an undergraduate program are imperative for propelling students into the field long-term, this is especially prominent for students from underrepresented minority groups [1,2]. Unfortunately, hands-on undergraduate experiences are limited to highly competitive Research Experience for Undergraduate (REU) programs at large, fully funded, universities. For small and 2-year undergraduate programs there has been a push for to find alternatives to help offer more hands-on experiences to students. The Vertically Integrated Projects (VIP) program is one way to accomplish this goal. VIP courses consist of students of various age, experience and majors working to tackle problems of external value that lasts over multiple semesters [3]. At Cooper Union, which is a primarily undergraduate institute, one area of the VIP program focuses on developing new inquiry-based learning (IBL) modules for translational biomaterials research, an underrepresented topic in the literature. Additionally, there is a need to develop future researchers in the area of biomaterials-based engineering education.

Through the VIP framework, a graduate student section was implemented. In this section, master's students learn about engineering education pedagogy and work alongside the professor to add to the field of engineering education. Furthermore, the VIP helps train master's students to be near-peer mentors to the undergraduate students in the VIP. The presence of a master's student serves as an invaluable alumni resource that is more personalized than just the professor alone. Additionally, the masters student develops their own leadership, teaching, and problem-solving skills that are beneficial for their thesis research and beyond. The efficacy of this framework is currently being evaluated in a long-term study using a modified version of the validated Undergraduate Research Student Self-Assessment (URSSA) survey.

Currently, the biomaterial IBL learning modules in development are: assessing variations in fatigue strength of corroded paperclips, understanding controlled release mechanics of soft contact lenses with food coloring, and modeling pH-dependent particle diffusion through capsules. All modules are designed with inexpensive and easy-to-acquire items such as food coloring, vinegar, paperclips, etc. to remove the financial burden accustomed with biomaterials research. Once completed, these modules will be incorporated into an undergraduate biomaterials class and K-12 STEM Outreach, both in-person and remote. To provide an equitable, deployable, and successful biomaterials IBL modules for all ages, genders, and minorities is at the core of our mission. This will enable more students to be exposed to biomaterials research at a younger age and improve the diversity of the field, which helps engender more creative ideas to solve translation biomaterials research problems.

References

[1] S. Olson, D.G. Riordan, Engage to Excel: Producing One Million Additional College Graduates with Degrees in Science, Technology, Engineering, and Mathematics. Report to the President, in, 2012.

[2] D. Lopatto, Science in Solution, 2010.

[3] R. Abler, T. Juhna, H. Kim, S.P.E. Marshall, M. Pardo, J. Sonnenberg-Klein, W.S. Percybrooks, Vertically Integrated Projects (VIP) Programs at International Institutions: Multidisciplinary Projects with Homes in Any Discipline, in: American Society of Engineering Education Annual Conference, Columbus, Ohio, 2017.

366h Increase Recycle to Reduce the Purge – a System to Improve Curriculum Retention

John Wagner, Amanda P. Malefyt, Arthur Hersel, Jacob R. Borden

Students' ability to remember concepts from one course to the next is shockingly low. This is evident when students are asked to recall information and more so when asked to apply concepts from previous courses. In fact, some students who successfully complete engineering programs, carry over little knowledge from one course to the next. The current educational system where both students and faculty are evaluated based on individual courses is an integral part of this problem. Little time or effort is spent addressing this course-to-course concept / knowledge loss. While there have been some proposals to address this issue, e.g. spiral curriculum or project-based learning, the current course-to-course linear structure of engineering programs is unlikely to change for most universities over the foreseeable future.

A novel review system is being developed that attempts to address this issue. This review system will fit into any academic program curriculum and consists of two major parts. This first part is a concept question distribution system, and the second part involves a student-centered mechanism to create those questions.

The distribution system is initiated by first selecting concept areas for review and then scheduling the timing and frequency that questions are to be distributed. The instructor can pick individual questions for a time slot and let the system randomly select questions for the undesignated time slots from the selected concepts. For each day of scheduled questions, the question in the first time slot is emailed to all the students in the class. When the student responds to the email, it takes them to the QRQuestion web site where the student receives feedback and additional questions for that day. If the student gets the question incorrect the system will ask the student that same question again after a few days have passed. If the student gets the question correct, the student will not receive that question again for at least a couple of weeks. After a student answers the question correctly a couple of times in a row, that question is retired from the students backlog and the student's individual score in that concept area improves. By keeping track of the performance in this way, a concept "heat map" can be developed for each student. The question delivery system incorporates repetition, spacing and interleaving. These techniques have all been shown to be important in long-term information retention. The instructor is emailed the questions and students performance on the questions. This makes it possible for the instructor to review the problematic questions in the next class and provide additional explanation.

It is important to have a mechanism to remove ambiguous or otherwise ineffective questions from the system. The first time a student sees a question, but before they answer it, they rate the question on clarity and relevance. Questions that receive low ratings or are too easy or too difficult as measured by student performance are eliminated from the question bank. In addition to a question being categorized according to concept, questions are also classified by the author as "Basic Knowledge", "Basic Concept", "More Advanced Concepts" or "Questions Involving Calculations". "Basic Concept" and "More Advanced Concepts" type questions provide an explanation and "Questions Involving Calculations" include worked out solution as part of the feedback.

The point system for student participation is set up to reduce the incentive for cheating. The first time a student sees a question, the points received are the same whether they answer it correctly or not. When students receive the question again, points are removed if they answer it incorrectly.

The second part of the review system is the question creation process. Questions can be input by the instructor or students can use the system to create questions and then review and edit questions submitted by other students. In this way, students interact with the course content at a deeper level than simply answering questions, aiding in content mastery by the student. Students receive points for both authoring and reviewing questions. Student generated questions that have been reviewed and approved by two other students, serve as candidate questions for entry into the distribution system mentioned above. Candidate questions are then reviewed by a subject matter expert, typically the instructor, and promoted, rejected, or sent back to the student author. If the question is rejected or sent back, the instructor can decide if the editors and/or author should lose points for errors that should have been caught.

The Chemical Engineering Program developing this system is ideal for evaluating the effect of this review system since the department has many years of data for several key performance indicators of student retention of program content. The curriculum assessment exam, CAE, was developed by the faculty of the department over 20 years ago. The CAE is a 6-hour 2-part multiple choice examination covering the required curriculum taken by ChE students in their final semester. The department has also required students to register for the NCEES Fundamentals of Engineering Examination as a graduation requirement for the last 4 years. In addition, an objective mathematics assessment examination has been taken by senior chemical engineering students for over a decade.

The review system will be implemented in designated required courses throughout the program so students are constantly reviewing previous course content from sophomore through the senior year. The combination of the two parts of the system should help students "recycle" previous course content to reduce content "purge" and provide a mechanism to monitor personalized student progression through our program. Aggregate improvements can be monitored longitudinally from the department's key performance indicators. If found effective, this system could serve as a template for other systems across other academic disciplines

366i Developing a Framework to Examine Women STEM Faculty's Participation in Entrepreneurship Education Programs

Aida López Ruiz, Prateek Shekhar, Jacqueline Handley, Aileen Huang-Saad

Over the past decade, there has been an increase in entrepreneurship education programs (EEPs) offered to faculty in universities, especially in science, technology, engineering, and mathematics (STEM) fields. Given the economic impact of research translation, EEPs have become an essential tool in academic settings to accelerate the translation of STEM research into marketable products. However, the underrepresentation of women in STEM entrepreneurship spaces is an ongoing problem that warrants further investigation. EEPs provide opportunities to address women's underrepresentation in the STEM entrepreneurship space by supporting women in entrepreneurial training and practice. Although several EEPs in university settings have been created, research examining women faculty's participation in EEPs is minimal. Furthermore, there is a need for a detailed, theoretical understanding of STEM women faculty participation in EEPs, to inform the development of inclusive and equitable programs.

In the presented study, we describe our project involving the development of a qualitative framework to build an understanding of women STEM faculty's participation/non-participation in EEPs. In-depth interviews of 32 self-identified women faculty, 19 non-EE participants and 13 EEP participants were used as the data source. The sample was racially and ethnically diverse, achieving similar representation across women identifying as Black, Latina, Asian and White. Furthermore, to incorporate different STEM disciplines, maximum variation sampling across engineering, natural science and mathematics was applied to recruit the participants. The research question examined was – What factors inform women STEM faculty's participation/non-participation in EEPs?

Qualitative analysis of the interview data using first and second cycle coding, identified factors that inform women faculty's participation/non-participation. Building on the previously reported Participation in Entrepreneurship Education Program (PEEP) Model, a qualitative framework was developed that further explicated the factors and sub-factors informing women STEM faculty's participation in EEPs that are noted in the pertinent literature. The results from this study provide guidance to administrators and practitioners engaging in university EEPs in developing new or revising ongoing entrepreneurship programming to create inclusive and equitable programming for women STEM faculty, and consequently promote diversity, equity, and inclusion in the area of STEM entrepreneurship.

366j Self-Efficacy in First-Year STEM Majors Correlates with Academic Progression

Robert Petrulis, Sona Gholezadeh, Edward Gatzke

A group of first-year STEM majors with high financial need participated in surveys considering a variety of factors including: self-efficacy, sense-of-belonging, identity, teamwork self-efficacy, community involvement, and overall college experience. Student surveys were conducted at the start of their first semester and near the end of their second semester during the 2020-2021 academic year. Initial results for this student cohort indicate that a net positive increase in self-efficacy appears to correlate with progression in the college to a second year. One student had a slight decrease in the mean efficacy value and decided to change to a technical major outside the College of Engineering and Computing. Another student had a significant decrease in the mean efficacy value and did not continue on at the University. Details of the survey and results from other factors are provided. Inference based on this survey is being used to help guide university efforts in support of first-year student retention.

366k Using Matlab, Simulink, and Simscape in Chemical Engineering Fluid Mechanics Courses

Aycan Hacioglu, Terry Denery, Bradley Horton

In chemical engineering, many chemicals are transported from one unit operation to another and processed in a fluid or partially fluid phase. To equip chemical engineers with the essential understanding of fluid mechanics, many chemical engineering programs list fluid mechanics or equivalent courses as core courses.

In typical undergraduate introductory fluid mechanics courses, fluid statics, control volume analysis (mass, momentum, and energy conservation), differential fluid flow analysis (continuity, Navier-Stokes Equation), the Bernoulli equation and its applications, flow in pipes (laminar, turbulent flow and frictional losses) are covered. In these courses, students usually learn analytical methods to solve fluid mechanics problems. This approach helps students master the fundamental concepts, but the problems that could be solved analytically are limited. In the workforce, students encounter more complex flow problems and use software to tackle such problems.

Students' understanding of these fluid mechanics concepts can be enhanced with additional exercises involving computational tools. MATLAB is a high-level technical computing language used for data analysis, numeric and symbolic computation. Simulink is a graphical environment where dynamic systems can be represented as graphical and functional blocks. Simscape is a physical modeling tool to model and simulate multidomain physical systems. These tools are widely used in the engineering industry and academia. In this talk, we will discuss how the chemical engineering fluid mechanics curriculum could be modernized with the addition of these tools. Using such computational tools help students get started with the typical workflows followed in the industry and expand the variety of problems students can solve. Through examples derived from typical chemical engineering fluid mechanics problems such as tank draining, constant head tanks, and pressure loss calculations, we will showcase how students can learn to use computational tools and master fluid mechanics concepts at the same time.

366l Ocular Drug Delivery and Unsteady-State Mass Transfer: A Project-Based Learning Approach Using Comsol Multiphysics

Daniel Lepek

Project-Based Learning is a pedagogical technique in which students complete specifically-.designed projects that aid students in their mastery and understanding of course material. While these projects may be hands-on in nature, software packages such as COMSOL Multiphysics have enabled the development of simulation-based projects. The simulation-based project described here is an unsteady-state mass transfer study regarding the ocular drug delivery characteristics associated with multi-layered ophthalmic (contact) lenses assigned to third-year undergraduate students in the Heat and Mass Transfer course offered at The Cooper Union.

Ocular drug delivery systems, such as a drug-loaded contact lenses, are used to obtain therapeutic effects for medical conditions of the eye. While contact lenses have commonly been used as an occular drug delivery vehicle, surface modifications, such as coatings, can be used to control the drug release behavior. A study by Pimenta, et al., (Diffusion-Based Design of Multi-Layered Ophthalmic Lenses for Controlled Drug Release) was used as the basis for a simulation-based COMSOL Multiphysics project that was designed to teach third-year undergraduate chemical engineering students the principles of unsteady-state mass transfer and diffusion [1]. While this topic is covered in the Heat and Mass Transfer course offered at The Cooper Union, associated analytical solutions and numerical methods techniques are not covered in this course. Therefore, COMSOL Multiphysics served as an ideal pedagogical vehicle to provide students with a engaging learning experience regarding this topic.

A detailed description of this COMSOL Mutiphysics project as well as representative graphical samples and solutions obtained from completed student work will be provided.

[1] Pimenta AFR, Serro AP, Paradiso P, Saramago B, Colaço R (2016) Diffusion-Based Design of Multi-Layered Ophthalmic Lenses for Controlled Drug Release. PLoS ONE 11(12): e0167728. doi:10.1371/journal.pone.0167728

366m Development of 3D Printed Desktop Learning Module for Learning Packed and Fluidized Bed Concepts

Zeynep Durak, Bernard Van Wie, David B. Thiessen

In previous years, our team worked on developing Low-Cost Desktop Learning Modules (LCDLMs) to create an interactive learning environment to enhance undergraduate STEM education. This year our team will be working on designing a desktop version of the fluidized bed using the 3D-printing methodology. The newly designed fluidized bed will be used to display phenomena to aid in conceptual understanding of packed bed and fluidized bed theory. The viscous loss term of the Ergun equation may be used to explain the relationship between pressure drop through the length of the bed as a function of such parameters as particle size, fluid velocity, length of the bed, fluid viscosity and density, and void space. Initially, the system will begin at a certain height as a packed bed with pressure drop increasing as superficial velocity increases. When the minimum fluidization point is reached, particles in the bed will expand upwards uniformly at a constant pressure drop across the bed. As superficial velocity increases further, the particles will again form a stationary packed bed, this time at the top of the column, where it is expected that the kinetic loss term of the Ergun equation will provide a significant contribution to the pressure drop. Glass beads are going to be used as packing material with water as the fluid medium. Classroom worksheets and pre- and post-tests are being developed to support the use of the fluidized bed in the classroom and assessment of conceptual gains that accompany this use.

366n A Hands-on Experience to Study Membrane Technology Developed By Undergraduate Chemical Engineering Students

Andie Veeder, Jackqueline Steinman-Ptacek, Natacha Souto-Melgar

Here we describe the development of an experiment to study separation using membrane technology. A reverse osmosis pilot plant equipment, donated by industry, was put in operational conditions at a low cost by a team of two undergraduate students. Our reverse osmosis system is simple, versatile, and portable. The undergraduate students improved the equipment design and prepared the operation and maintenance procedures. Also, they prepared an experimental handout and a video demonstration for the reverse osmosis experiment. This new experiment will be incorporated into the Chemical Engineering Senior Laboratory course. The students who participated in this project were exposed to a learning experience via an open-ended approach; they acquired the ability to develop and conduct appropriate experimentation, analyze, and interpret data, use engineering judgment to draw conclusions, and improve their technical writing and communication skills. This project has a broader impact, after implementing the reverse osmosis experience in the laboratory course, chemical engineering undergraduate students will have the opportunity to perform various hands-on experiments, learning the theory and practice of an important membrane separation process.

547a Capstone Design with Practical Characteristics for a New Era

Gavin Towler

The capstone design class serves the needs of a wide range of stakeholders. Students need to experience working on a design project where they can see how the academic concepts from foundation classes can be brought to bear on a large realistic engineering problem. They also need to learn enough about economic analysis, optimization and equipment specification to be competent and confident in entry level roles in industry. In many schools the senior design course also fills in various other gaps in the students' knowledge to better prepare them to enter the workforce. Educators often struggle to balance these competing needs. Many schools use Adjunct faculty to create a more authentic experience for the students, but at the risk of teaching design techniques that may have been best practice over forty years ago.

Design is intrinsically a constrained optimization problem, and so is the construction of a well-balanced capstone course that effectively meets the needs of students (regardless of their intended career path), employers (regardless of their sector of industry or commerce) and broader stakeholders such as ABET, professional engineering societies and the general public. Some things that are easily learned in a University are much harder to learn on the job and vice versa, and instructors need a good feel for the relative importance of topics and experiences to ensure they have optimized time allocation, project selection and student outcomes. I will share how my experiences as an instructor, adjunct advisor, textbook author and employer helped shape my formulation of the problem. As always in design, there is no single right answer!

547b How We Teach: Capstone Design Survey Results

Laura Ford, Janie Brennan, Jennifer Cole, Marnie Jamieson, Kevin Dahm, David Silverstein, Troy Vogel

The Curriculum Committee of AIChE's Education Division will survey chemical engineering departments in the US and Canada during September 2022 about capstone design courses. This talk will present results about class assessments, topics, and resources.

437a Becoming a Well-Rounded Scholar: A Spotlight on Effective Teaching and Mentoring

Allison Godwin

Both teaching and mentoring require skills that are often not taught explicitly and require in-the-moment assessment and adjustment to be most effective. In this panel, I will share tips and strategies centered in belonging-centered instruction and mentoring from evidence-based practices and my own experiences. These topics could include alignment of course content, pedagogy, and assessment of how to manage and lead an effective research group. Overall, this discussion will provide ways to shape your development as a scholar beyond research and help make all the aspects of your role effective.

437b Staying True to Your Vision and Values As a Junior Faculty

Karthish Manthiram

In this talk, I will share tips and strategies for successfully pursuing, evolving, and refining the vision that brought you to academia. This will intersect with topics regarding group management, including how to empower group members to feel ownership over the way in which the group is run and how to experiment not just in your science, but in the operation of a group. Overall, the strategies outlined here will provide a path to not just advancing your scientific vision, but staying true to the values you hold inside and outside academia.

107a The World Is Changing and We Must Change with It - Lessons Learned in Alternative Grading Approaches in Core Chemical Engineering Courses

Daniel Burkey, Jennifer Pascal, Matthew Stuber, Kristina Wagstrom

The COVID-19 pandemic brought significant disruption to the way that we deliver classes to students. Remote learning, absences from class due to personal or family illness, and other disruptive events have in many cases changed the desirability and practicality of traditional deadlines, assignments, and major assessments. In this presentation, four faculty members teaching core chemical engineering courses in the second, third, and fourth year discuss the design and implementation of alternative evaluation schemes in their classes over the course of the past two years, including challenges, lessons learned, and student response to these alternative schemes.

In the second year, we describe a specifications grading approach in a material and energy balances course built around individual projects in lieu of exams, opportunities for growth through resubmission, and use of reflections to develop better work habits. This shifted the goal to providing students the opportunity to demonstrate their understanding of core concepts rather than high-stress testing that evaluates their ability to demonstrate these concepts under the pressure of time.

In the third year, we discuss two courses, a Chemical Engineering Mathematical Analysis Course, where we utilize an augmented specifications grading approach whereby conventional assessments: homework problem sets, daily quizzes, and exams are offered to students in the standard way, but credit is earned based on a method that is less punitive and extremely forgiving of simple mistakes. Summarily, students are provided a 'choose your own adventure' approach to achieving a desired grade, with the lowest passing grade corresponding to the minimum acceptable achievement of the student outcomes. This approach was designed to be adapted to any existing class with standard assessment and grading formats with little redesign required. In Fluid Mechanics, an approach based on "Ungrading" was used, decentering grades while emphasizing feedback with a semester-long project designed to give students voice and choice in how they demonstrated their learning.

In the fourth year, we also applied an "Ungrading" approach to the Capstone Design course, which included the elimination of all intermediate deadlines, a "no-grades, feedback-only" approach to assignments, flexibility and choice in assignment completion, and a comprehensive portfolio with one-on-one student meetings to demonstrate and discuss learning outcomes.

107b Implementation of Achievement-Based Grading in Chemical Engineering Core Classes

Adam K. Ekenseair

Achievement-Based Grading (also similarly referred to as Specifications Grading or Mastery Grading) is an assessment structure where students work towards demonstration of learning and skills and a final grade is awarded based on the number and level of achievements successfully completed. While implementations are varied, common themes include distinction between simple and more complex assessments, significant flexibility in when (and often in what order) learning and skills can be demonstrated, and opportunities to grow through failure by re-taking unsuccessful assessments. Some other aspects that may be found in particular implementations include pathways to the highest grade available without having to show achievement in all skills covered in the course and the ability to demonstrate achievement though a variety of means.

This assessment structure has appealed to educators for a number of reasons. One is the potential to streamline grading and make delivery of challenging material more manageable, though this ultimately depends on the particular implementation. Some more generally applicable benefits are associated primarily with a shift in student perceptions of what assessment means and how progress through curricular topics is accomplished. The focus is placed on the topical material and course outcomes and students' ability to demonstrate those outcomes by the end of the term. Instead of worrying (or perhaps complaining) about how points were lost on one assessment or another, students instead have a path forward where their grade is earned based on demonstrated achievements and where they are encouraged to revisit topics until mastery can be shown (which can cut down on the prevalence of cheating). This structure also facilitates a straightforward means of directly challenging the average student to achieve a higher degree of learning in a course. Furthermore, instructors can readily project a supporting role in the progression of students' development. Instead of feeling like grades were unfairly or even arbitrarily determined, students can see the expectations from the beginning and relate their final grade to their own accomplishments, or lack thereof.

A variation of Achievement-Based Grading has been successfully implemented in a Chemical Engineering Thermodynamics 1 course (entering its fourth iteration in Fall 2022) and is being implemented for the first time in a Materials and Energy Balances course in Fall 2022. The details and evolution of this implementation, lessons learned, student feedback, and impacts on student learning will be discussed.

107c Perspectives on Alternative Assessment in Core Chemical Engineering Courses

Joshua Enszer

The switch to fully-online courses during the COVID-19 pandemic required a rethinking of how we measure attainment of student learning outcomes, as we traditionally relied on in-class exams or long-form reports as the primary modes of assessment in our lecture- and lab-based courses, respectively. In several core courses at our university, we replaced these traditional assessments with alternatives. In lower-level courses, we restructured and rebranded quizzes and exams as "checkpoints," regularly spaced across the semester, which comprised of some typical homework and exam questions but included extensions to these questions, such as identification and reflection related to specific learning outcomes. In our senior-level precursor to capstone design, we implemented portfolios, in which students created their own prompts and solutions to demonstrate their learning. In one semester of our laboratory sequence, we used the opportunity to reflect anew on course goals and replaced a good portion of our traditional writing assignments with live "in-person" meetings, with the added benefit of giving students more practice writing in different formats (such as short reports and memos) as well as running their own meetings with faculty, creating agendas and minutes. Each of our changes in assessment had their trade-offs, primarily in terms of time required: to communicate expectations to students working in these modes for the first time, as well as to provide accurate grades and meaningful feedback.

Checkpoints were used to replace traditional timed in-class exams for both our material and energy balance and engineering statistics courses. These comprised of a short "same-day" component (meant to be completed within 24 hours of posting to our learning management system) followed by a "long-term" component (meant to be completed in 3-5 days, in place of a weekly homework assignment). The long-term component included additional prompts to identify which learning outcomes were demonstrated by completing the technical problems, and identifying other problems from class, reading, or homework that helped students feel confident in their work. These reflective components helped to ensure some level of academic integrity – no two students would submit the same reflection – though it is unclear whether or not this prevented students discussing the other problems during these "take-home" activities. We have resumed in-person classes but still use a modification to this checkpoint system rather than return to fully in-class timed exams. Now we hold a take-home exam over the span of 4-6 days, and the day the exam is due is followed by a short in-class quiz, with the intention that preparing well for the exam serves as good preparation for the in-class quiz. So far there is generally good correlation between the two components of a checkpoint, and we have substantially fewer students requesting accommodations (extra time on tests, etc) because of the flexibility of the take-home exam and focus of the in-class quiz. Grading checkpoints takes no more or less faculty time as compared to traditional exams.

In our senior lab course, historically teams of students prepare lengthy (25+ pages) reports on unit operation labs. In our fully-online version, these reports were broken into small components, requiring regular meetings in Zoom with course faculty. For example, the distillation lab documentation was replaced with three smaller writing assignments, each of which related to one another, rather than just revisions and lengthening compared to previous years. Students also gained practice with many more forms of written communication – memo and short-report formats, communication of agendas and meeting minutes, and preparing things other than PowerPoint slides to present and meet with course faculty. Faculty grading time was broken up into smaller, often contemporaneous grading during or shortly after meetings, requiring a different kind of attention compared to reading lengthy one-off reports.

In our first semester of our two-semester senior design sequence, traditional exams on equipment sizing and sequencing, economics, and process safety were replaced with portfolio assignments. In a portfolio, each student created 2-3 problems of their own together with sample solutions, given a list of learning outcomes and constraints as to how the work should be presented. Students reported feeling much more confident about their learning, and scores on portfolios were generally very good, in agreement with student perceptions. However, this form of assessment is much more substantial than a traditional exam, so care needs to be taken in defining constraints and rubrics to help faculty use their time well in scoring and providing feedback.

107d Ungrading in Process Control: Attempting to Eliminate Exams, Deadlines, and Anxiety By Refocusing on Learning Instead of Grades

Lucas Landherr

Modern means of assessment require approaches that consider fair and equitable means of determining students' understanding and performance. For example, students frequently connect exams to stress and anxiety, and thus believe them to provide inaccurate representation of their ability level. On a broader level, centering the focus of a course on what is necessary to achieve the highest grade drives students to focus more on their numerical scores than on the depth of their learning; effectively, the course outcomes become achieving a grade instead of student understanding.

Ungrading provides an approach that shifts the emphasis of each course back onto learning and what students should be able to do at the end of a course. Grades are de-emphasized in exchange for greater levels of discussion and feedback focused on how well students have learned, processed, and applied the instruction in the course. There are many approaches to ungrading such as contract grading, but the broader approach allows an instructor to provide less stressful, more equitable assessment.

In a process control course, an ungrading method was applied that eliminated exams and most deadlines in the course. Students were provided a number of optional problems and exercises that could be conducted that aligned with each of 14 course outcomes. Exercises could be submitted for feedback throughout the semester, allowing students to correct their work and assemble a completed portfolio of work demonstrating their mastery over the course outcomes at the end of the semester. Some course outcomes could be completed by work on a semester-long course project, while other exercises were simply homework and exam problems from previous iterations of the course restructured to allow students to process their understanding and better apply their skills for a more considered performance of understanding. Students completed a number of self-assessment assignments throughout the semester, and a final grade was determined for each student in discussion with the instructor based on a reasoned consideration of their efforts and completed correct work.

This paper will discuss these ungrading efforts in the course, student feedback throughout the semester, and recommendations for other instructors interested in applying an ungrading approach in their courses.

107e Efficacy and Perceptions of a Specifications Grading Scheme for Chemical Engineering Seniors

David Wagner

The purpose of this research is to determine the efficacy of using a 'specifications grading' scheme in the senior year of undergraduate education. It could be that the switch to such a system is beneficial or detrimental to students since they are likely used to a point-based (0-100 points) grading scale. The survey of chemical engineering seniors acquired information of and perception on a point-based versus specifications grading scheme. Studies on the efficacy of grading and correlations to assessment outcomes are ongoing in academia [1]. Entire disciplines focus on effective instruction and the psychology of student fulfillment when it comes to school and subsequent graduation. Standards-based grading [2] and specifications grading schema are closely tied together when the focus shifts from an instructor-centric lecture to a student-centric one. Instructors have seen improvements in grades and behaviors toward grading in chemistry [3] and writing courses [4].

A similar approach was taken as in references Howitz, et al. [3] and McKnelly et al. [4], in which the course switched from a point-based system to one that uses specifications grading and standards. Final grades still utilize the A, B, C, D, and F letter grades. I incorporated many of the same aspects as other researchers to show either consistency or a lack of when it comes to reward systems, resubmission, and improved mental health regarding grades (overall).

Participants were surveyed via Google Forms in January 2022, with a follow up of the same survey in May of 2022 to investigate how perceptions change over the semester. They were surveyed online and outside of class time. The survey has an introduction detailing anonymity of the part of the students and describes the survey as voluntary. Overall grade improvement is summarized with key take-aways regarding the efficacy of the grading scheme and recommendations for further improvements. Adaptation strategies are also discussed for those wishing to make any switch in their own courses with a summary of best practices found in the literature.

References

[1] M.Y. Tsoi, M. Anzovino, A. Erickson, E. Forringer, E. Henary, A. Lively, M. Morton, K. Perell-Gerson, S. Perrine, O. Villanueva, M. Whitney, C. Woodbridge, Variations in Implementation of Specifications Grading in STEM Courses, Georg. J. Sci. 77 (2019). https://digitalcommons.gaacademy.org/gjs/vol77/iss2/10 (accessed January 10, 2022).

[2] J. Elsinger, D. Lewis, Applying a Standards-Based Grading Framework Across Lower Level Mathematics Courses, Https://Doi.Org/10.1080/10511970.2019.1674430. 30 (2019) 885–907. https://doi.org/10.1080/10511970.2019.1674430.

[3] W.J. Howitz, K.J. McKnelly, R.D. Link, Developing and Implementing a Specifications Grading System in an Organic Chemistry Laboratory Course, J. Chem. Educ. 98 (2021) 385–394. https://doi.org/10.1021/ACS.JCHEMED.0C00450/SUPPL_FILE/ED0C00450_SI_002.XLSX.

[4] K.J. McKnelly, M.A. Morris, S.A. Mang, Redesigning a "writing for Chemists" Course Using Specifications Grading, J. Chem. Educ. 98 (2021) 1201–1207. https://doi.org/10.1021/ACS.JCHEMED.0C00859/SUPPL_FILE/ED0C00859_SI_003.DOCX.

107f Assessing Authentic Problem-Solving in Heat Transfer

Jiamin Zhang, Soheil Fatehiboroujeni, Matthew Ford, Eric Burkholder

Problem-solving is an essential skill for engineers-in-training to develop. Although engineering programs stress the importance of teaching problem-solving skills, there are frequently reported gaps between the skills graduating engineers have and what employers want. Part of the reason for this gap is that many of the problems students solve, particularly in engineering science courses like thermodynamics and fluid mechanics, bear little resemblance to the authentic, unstructured problems they will be expected to solve as engineers. Recent work has characterized authentic problem-solving as a set of 29 decisions common across experts in different STEM fields. These decisions represent specific, measurable skills that can be practiced with appropriate feedback to develop problem-solving expertise. In this paper, we describe the development of an assessment to measure decision-making in the context of heat transfer. This assessment represents a departure from typical assessments in that it is not designed to rank students relative to one another, but to measure the progression of students' expertise toward a standard established by a consensus of experts. We tested the assessment with 14 engineers highly skilled in heat transfer and transport phenomena, as well as 12 undergraduate engineering students who had previously taken a heat transfer course. From this data, we developed a scoring rubric that maximizes the discrimination between expert thinking and student thinking. On many decisions, students are far from expert-like, whereas for others, students are much closer to expert thinking. In some areas where expert responses differ from one another, our think-aloud interviews reveal how the specific decisions made by experts depend on their interpretations of the goals in an unstructured problem-solving context. We hope to encourage other educators to use this assessment in their courses to measure how well they are preparing their students to solve real-world engineering problems.

107g Assessment of an Inquiry-Based Learning Bioadhesive Module to Teach Principles of Tissue Repair

Chris Panebianco, Poorna Dutta, Jillian Frost, Angela Huang, Olivia Kim, James Iatridis, Andrea Vernengo, Jennifer Weiser

Introduction: Biomaterials have had an enormous impact on how diseases are understood, diagnosed, and treated since their inception in the 1940's [1]. Bioadhesives, which are designed to adhere biological components together, are an important class of biomaterials for surgical wound healing and repairing soft tissue defects [2,3]. To develop the next generation of bioadhesives, there is a societal need to teach trainees about their design, engineering, and testing. This study designed, implemented, and evaluated a hands-on, inquiry-based learning (IBL) bioadhesives module to teach undergraduate, master's, and PhD/postdoctoral trainees about bioadhesives. The goal for this IBL bioadhesives module was to promote positive learning gains and enhance scientific literacy for a wide-range of trainees with limited knowledge of bioadhesives.

Methods: Instructors guided trainees through this 3 hr IBL module at The Cooper Union for the Advancement of Science and Art (undergraduate cohort), ETH Zürich (master's cohort), and the Baltic Biomaterials Centre of Excellence (PhD/postdoctoral cohort). The first part of this module was a pre-laboratory exercise, which consisted of a brief background on bioadhesives and trainee-led discussions about experimental design. Next, trainees adhered two different substrates (*i.e.*, chamois and 3D-printed polylactic acid (PLA)) together using two different bioadhesive (*i.e.*, glutaraldehyde (GTA)-crosslinked gelatin and cyanoacrylate). GTA-crosslinked gelatin and cyanoacrylate were chosen as bioadhesives because they adhere through mechanical and chemical mechanisms, respectively, and because their components are commonly used for tissue repair applications [4]. Trainees tested the adhesive strength of their bioadhesives using a lap-shear configuration that mimics ASTM 2255-05, an internationally recognized protocol for testing the strength properties of tissue adhesives [5]. Briefly, trainees attached one side of their substrate to a digital luggage scale placed on top of a ring stand, then pulled the other side of their substrate to failure while recording the force reading with a smartphone. After completing this module, trainees completed a post-laboratory report or exercise, which introduced bioadhesives, described tested independent variables (*e.g.*, GTA concentration, chamois vs. PLA), summarized results with appropriate statistical analyses and discussed findings in the context of tissue repair.

To assess the learning outcomes of these experiments, pre- and post-tests and pre- and post-surveys were given to each cohort. Pre/post-tests questions were designed by the authors to test trainee understanding of design variables and applications of bioadhesives for tissue repair. Pre/post-survey questions came from the Scientific Literacy and Student Value in Inquiry-guided Lab Survey (SLIGS) to assess scientific literacy [6]. A paired Student's t-test ($p < 0.05$) was used to compare the average pre-/post-test and pre/post-survey scores. All questions were approved by The Cooper Union for the Advancment of Science and Art Institutional Review Board (IRB), the Ethics Commission of ETH Zürich, and the Baltic Biomaterials Centre of Excellence.

Results: Twenty-nine trainees (N=29) completed the pre/post-test assessment. Overall, pre/post-test results showed that trainees experienced highly significant learning gains from completing this IBL bioadhesives module (Figure 1A). Follow-up analyses of individual cohorts showed that undergraduate (N=10), master's (N=9), and PhD/postdoctoral (N=10) cohorts all demonstrated equivalently significant learning gains (Figure 1C). While not significantly greater, the undergraduate cohort experienced the largest learning gains, with an average increase of 34.2 points.

Twenty-nine trainees (N=29) also completed the pre/post-survey assessment. Overall, results demonstrated that trainees were more confident posing scientific questions (SQ1), assessing methodology (SQ2), providing scientific explanations (SQ3), designing experiments (SQ4), and relating scientific discovery to societal impact (SQ5) as a result of completing this IBL bioadhesives module (Figure 1B). Similar to the pre/post-test assessments, follow-up analyses demonstrated that the undergraduate cohort (N=10) experienced the greatest improvements in scientific literacy (Figure 1D). Trainees in the master's (N=9) and PhD/postdoctoral (N=10) cohorts only experienced trending increases in some categories of scientific literacy.

Discussion: This study designed, implemented, and evaluated an IBL bioadhesives module that effectively taught principles of bioadhesives to undergraduate, master's, and PhD/postdoctoral trainees. This IBL bioadhesives module was implemented at multiple education levels because it was hypothesized that it could be effective for a broad range of trainees. For example, trainees in the undergraduate cohort experienced the greatest learning gains, which was expected since they would have had the least theoretical and applied knowledge of bioadhesives. Such learning gains are consistent with those in other published biomaterials experiments [7, 8]. More advanced trainees in the PhD/postdoctoral cohort also experienced significant learning gains by pre/post-test assessment. While PhD/postdoctoral trainees had more experience with scientific inquiry, their field of study materials science and chemical technologies and they had limited exposure to bioadhesive hydrogels. Therefore, this IBL bioadhesives study effectively introduced trainees in this cohort principles of bioadhesives, which they could incorporate into their primary area of research. These slightly reduced learning gains were consistent with biomedical engineering modules deployed on senior engineering students [8].

This IBL bioadhesives module also broadly enhanced trainee scientific literacy. The pre/post-survey used to assess scientific literacy employed validated survey questions to avoid self-reporting bias [9]. Notably, upon follow-up analyses, increases in scientific literacy were only significant for trainees in the undergraduate cohort. Trainees in the undergraduate cohort would have had the least exposure to scientific inquiry; thus, it is understandable that they would have had the greatest benefits. Moreover, since trainees in the master's and PhD/postdoctoral cohorts had significant experience leading scientific investigations, it was expected that they would have less significant improvements in scientific literacy.

References:

[1] Biomaterials [Internet]. [cited 2022 Mar 11]. Available from: https://www.nibib.nih.gov/science-education/science-topics/biomaterials

[2] Tarafder S, Park GY, Felix J, Lee CH. Bioadhesives for musculoskeletal tissue regeneration. Acta Biomater. 2020 Nov;117:77–92.

[3]. Li J, Yu X, Martinez EE, Zhu J, Wang T, Shi S, et al. Emerging Biopolymer-Based Bioadhesives. Macromol Biosci. 2022 Feb;22(2):e2100340.

[4] DiStefano TJ, Shmukler JO, Danias G, Iatridis JC. The Functional Role of Interface Tissue Engineering in Annulus Fibrosus Repair: Bridging Mechanisms of Hydrogel Integration with Regenerative Outcomes. ACS Biomater Sci Eng. 2020 Dec 14;6(12):6556–86.

[5] Standard Test Method for Strength Properties of Tissue Adhesives in Lap-Shear by Tension Loading. ASTM International. 2005 Mar;

[6] Ankeny C, Stabenfeldt S. Cost-effective, Inquiry-guided Introductory Biomaterials Laboratory for Undergraduates. 2015 ASEE Annual Conference and Exposition Proceedings. ASEE Conferences; 2015. p. 26.412.1-26.412.16.

[7] Farrell S, Vernengo J. A challenge-based laboratory to explore drug delivery from swellable matrices. Education for Chemical Engineers. 2016 Oct;17:21–30.

[8] Vernengo J, Dahm KD. Two challenge-based laboratories for introducing undergraduate students to biomaterials. Education for Chemical Engineers. 2012 Jan;7(1):e14–21.

[9] Howard GS. Response-Shift Bias. Eval Rev. 1980 Feb;4(1):93–106.

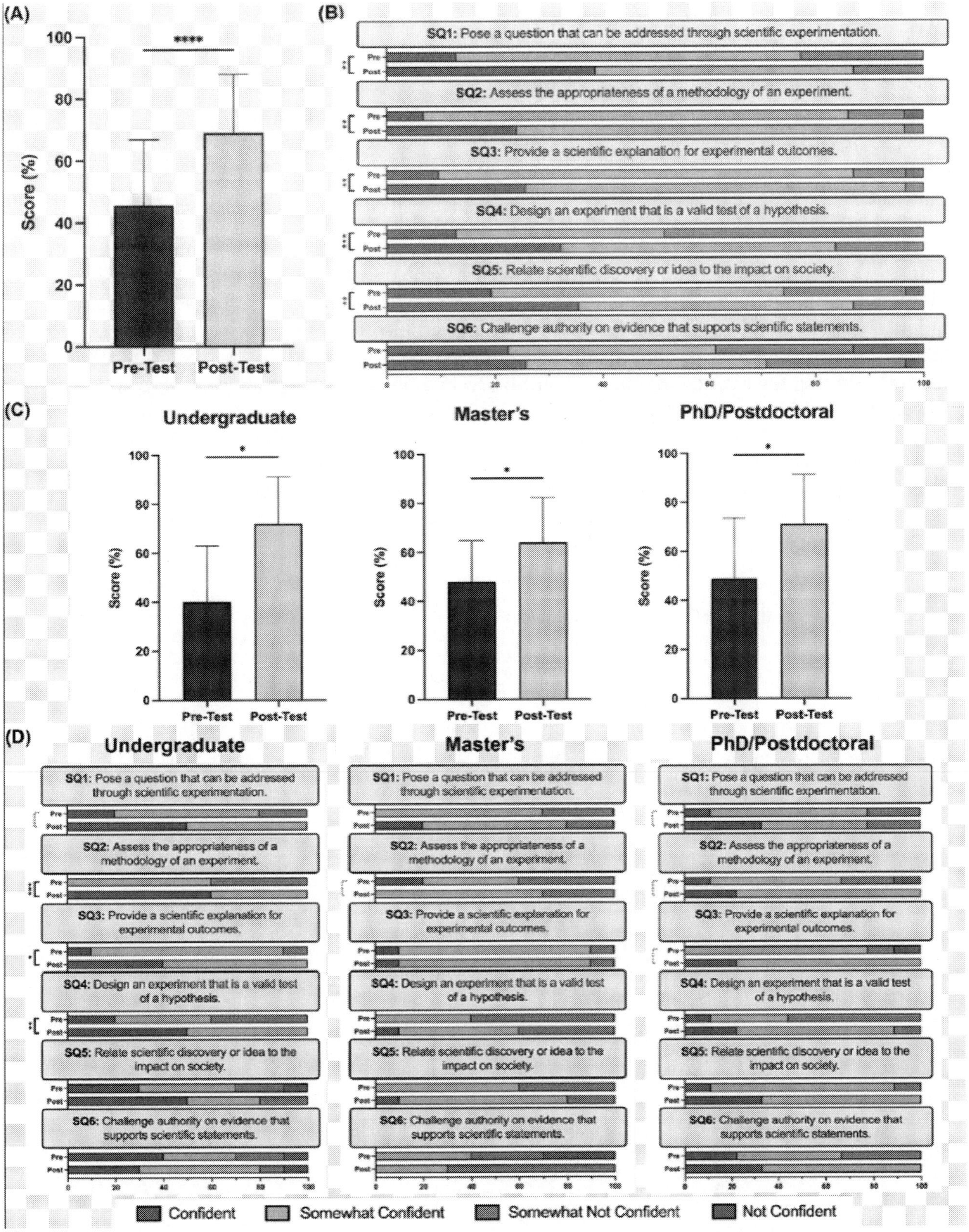

(A)
****
Score (%)
Pre-Test
Post-Test
(B)
SQ1: Pose a question that can be addressed through scientific experimentation.
SQ2: Assess the appropriateness of a methodology of an experiment.
SQ3: Provide a scientific explanation for experimental outcomes.
SQ4: Design an experiment that is a valid test of a hypothesis.
SQ5: Relate scientific discovery or idea to the impact on society.
SQ6: Challenge authority on evidence that supports scientific statements.
(C)
Undergraduate
Master's
PhD/Postdoctoral
(D)
Undergraduate
Master's
PhD/Postdoctoral
Confident
Somewhat Confident
Somewhat Not Confident
Not Confident

107h Comparing Formative and Summative Auto-Graded Problems for a Material and Energy Balances Course

Samantha Yanosko, Matthew Liberatore

Digital tools providing students with immediate feedback are well aligned with many best practices related to learning. Specifically, auto-graded, randomized problems for a Material and Energy Balances course has quantified correct and attempts for over 100,000 problems over more than 5 cohorts. These "big data" were obtained with using the fully interactive online textbook titled the Material and Energy Balances zyBook. In general, a median correct over 90% has been presented in previous work with median attempts before correct less than 3 for almost all problems. Thus, with unlimited attempts before weekly due dates, students are generally successful in solving auto-graded problems. On one hand, all cohorts of interest have used in-section, formative auto-graded problems that apply scaffolding, i.e., moving from easier to more difficult problems within a set of 3 to 7 problems. On the other hand, summative, end-of-chapter auto-graded problems were added and analyzed for three recent cohorts. Summative problems include both traditional problems as well as student-written YouTube problems, which has been discussed and published in recent years. Research questions included: 1. Do students correctly solve formative problems more readily than summative problems?, 2. Do formative and summative auto-graded problems correlate differently with final course grades?, and 3. Can too many problems be assigned? Some discussion on variation and reproducibility from cohort to cohort will also be discussed.

108a Computing and Data Analytics through Problem Based Learning

Weiguo Xie, Richard Davis

The requirement from employers on the skills of computing and data analytics for newly hired Chemical Engineering graduates is ever rising. As our industrial advisor boards also recommend, we continue enhancing our students' learning of computational methods, modeling, simulation of chemical engineering processes, and data-driven process optimization. We initially required one introductory programming course taught by the Computer Science department, an undergraduate course in numerical methods that included an introduction to chemical process simulation, and a graduate chemical engineering analysis course that included advanced simulation. Through our process of continuous improvement, we found that the introductory programming course was ineffective and that our fourth-year students lacked confidence and skill with process simulators. To address these issues, we dropped the computer science class, added coding into the computational methods course, and added a new course in process modeling and simulation. We also updated our advanced Analysis course with topics in data science. Computational Methods and Process Modeling are required core courses in our undergraduate program. Our graduate program requires the Analysis course, and undergraduate students may take this course to fulfill a technical elective requirement.

We have taught Excel with Visual Basic for Applications (VBA) in the Computational Methods course since 2010 and added Python for scientific computing and data analytics in 2019. Process Modeling uses Aspen HYSYS and SolidWorks. MATLAB was selected as the computational language for the advanced Analysis course in 2020. These courses aim to train our chemical engineering students to higher proficiency with modeling and simulation methods and computer software tools.

Process Modeling is a computer lab-based course to train students in using commercial simulation and modeling tools to prepare them for our unit operations and design courses. In our Computational Methods course, students learn topics including Introduction to Numerical Methods, Excel for engineering analysis, Pivot Table and Charts, VBA and Python programming, Linear equations, Derivative approximations, Roots of nonlinear equations, Optimization, Uncertainty, Least-square regression, Interpolation, Integrals, Initial-value problems and ODEs, Boundary-value problems, and PDEs. In Analysis, students learn topics including Introduction to modeling in chemical engineering, MATLAB skill preparation, Model fitting, spline functions, Optimization toolbox, Enhanced optimization, Global optimization, Optimal experimental design, Data statistics and analytics, Complex equation systems, Process integration, and optimization.

Apart from regular assignments and exams, the teaching pedagogy of Problem Based Learning (PBL) is used in our courses. In Computational Methods, we assign a term project that requires a combination of numerical methods. This assignment addresses our program student outcomes of complex problem solving and technical communication. For example, a project may require students to build a packed-bed reactor simulator in Excel. The reactor model equations involve mass, energy, and pressure balances requiring numerical integration of the first-order differential equations. Students also code functions for physical and chemical properties that require least-squares regression, numerical quadrature, or root-finding. Figure 1 shows an example of an Excel worksheet with a graphical user interface and plots of the packed-bed reactor profiles for composition, temperature, and pressure. Students are expected to incorporate data input validation and write a concise instructional manual for their simulation tool. Students may elect to prepare a video user guide.

We apply PBL for the students' final assignment in the Analysis course. They apply an orthogonal experimental design to find the optimal parameters that minimize the operation costs in solvent extraction flowsheet using a ternary diagram shown in Figure 2. Students have 3 weeks to work in teams and submit a written group report for this assignment.

In engineering education, interpersonal skills, teamwork, communication, and problem-solving skills are most frequently identified as highly important by employers. The PBL is an effective method for teaching computing and data analytics in these courses. The key features of PBL are to provide professional training involving both teamwork and individual exercises for our students and encourage a transformative learning experience for them.

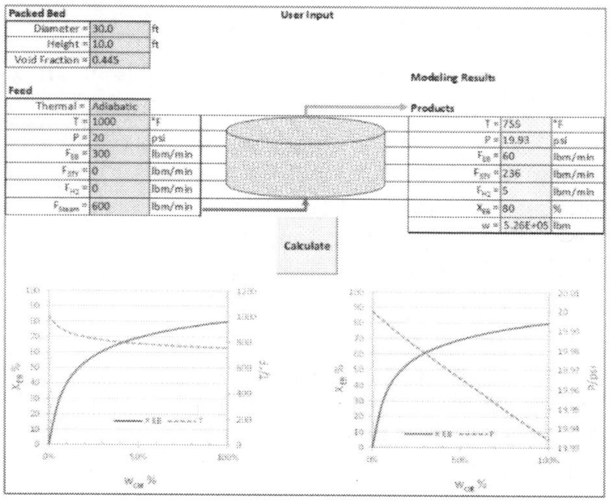

Figure 1 *Graphical user interface for a packed bed reactor simulator in Excel with temperature and pressure profiles.*

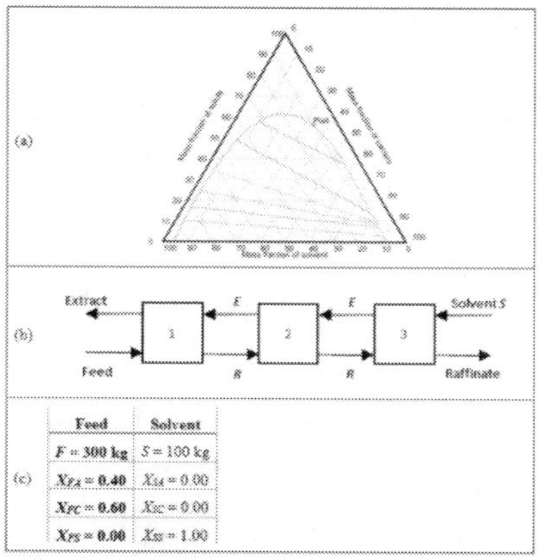

Figure 2 *Final assignment of CHE5031 to apply the OED method for finding optimal parameters that minimize the operation costs in solvent extraction flowsheet (a) The ternary phase diagram; (b) Counter-current three-stage solvent extraction circuit; (c) the feed and solvent mass rates and fractions.*

108b Teaching Process Simulation Concepts in Aspen Plus Using the Ammonia Synthesis Loop

Jason White

Formative process modeling assignments can greatly enhance students' confidence and competence in effectively using process modeling software prior to the capstone design project. The ammonia synthesis loop is a relevant and simple process with three components that can be used to demonstrate setting up reactor models and recycle loops using process modeling software. The ammonia process includes a reactor (PFR w/ catalyst), two heat exchangers, a compressor, and a vapor-liquid separation.

In order to increase student confidence and competence using Aspen Plus, students in an introductory process design course were assigned a 2-week project in which they were tasked to produce a simulation of the ammonia synthesis loop starting from pure nitrogen and hydrogen reactants. Since the single pass conversion of nitrogen and hydrogen to ammonia is low, recovering and recycling unused reactants is critical, challenging students to set up a recycle stream. Through this assignment, which was proceeded by a module on "Aspen Basics", students would be independently building a process model in Aspen Plus for the first time.

Regarding the reactor modeling part of the assignment, students gained experience in setting up a Langmuir-Hinshelwood-Hougen-Watson (LHHW) kinetic model for the ammonia synthesis reaction. They modeled the packed bed reactor using a PFR model with catalyst. Tutorials were included for how to input the LHHW rate expression. One of the deliverables for the assignment was to report on the reactor dimensions that they would prescribe using heuristics and the component composition profiles along the length of the reactor.

One of the most valuable aspects of this assignment was that students got experience in building the recycle structure of their flowsheets. Setting up recycle streams in Aspen can be tricky, and Aspen often yields convergence errors due to components building up in (or being depleted from) the process. To help illustrate and navigate these potential mass balance convergence issues, students were presented with a "simulated storage model", which when integrated into their process model could help identify the amount of raw materials needed to be fed into the process when in continuous operation in order to meet a target production goal. Additionally, this "simulated storage model" could help identify inadequacy of separations leading to holdup of components in the recycle loop, as well as any components that may need to be purged from the process. Ultimately, this simulated storage model was to be deleted and replaced with a mixer (to combine the feed and recycle stream) and a purge stream if necessary.

As a part of the students' report on their process modeling efforts, they were expected to present:

1) the flow rate, composition, temperature, and pressure of the ammonia product (target flow rate was defined and varied for each student, and the purity specification for the product was > 99.5 wt% ammonia) ,

2) the flow rate, composition, temperature, and pressure of the hydrogen and nitrogen inputs,

3) the energy requirements of each unit operation, and

4) the length of the reactor and how they arrived at this result.

A direct assessment of students' ability to meet expectations with regards to these deliverables, as well as an indirect assessment of outcomes by surveying students on the impact of the assignment, was used to gauge the success of the assignment in helping students to be more confident and competent using Aspen Plus as a process modeling tool.

108c Deliberate Practice of Spreadsheet Skills Using Randomized, Auto-Graded Problems

Luke Gorbett, Kayla Chapman, Matthew Liberatore

Spreadsheets are a foundational tool for engineers, engineering students, and many others. Spreadsheet tools, including Microsoft Excel and Google Sheets, have many common formulas, functions, and other tasks across versions and platforms. Here, an interactive textbook from zyBooks provides learners the opportunity to practice spreadsheet skills. On one hand, content is delivered through interactive components that include animations as well as multiple choice and matching questions. On the other hand, executing spreadsheet skills using 100+ auto-graded, randomized questions examines the deliberate practice framework. Using this type of online homework, fraction correct regardless of attempts, number of attempts before correct, and number of attempts after correct provided metrics across several cohorts. Content was grouped as General spreadsheet skills, Functions, or Advanced skills. General spreadsheet skills and Functions generally showed greater median correct, ranging from 76 to 90%, than Advanced spreadsheet skills, which led to median correct of 68 to 81%. Median correct also varied by question types and order, which aligned the expert author's intent to scaffold the questions. Finally, a hypothesis was tested: Adding a Copy sheet feature to the auto-graded questions would encourage deliberate practice and improve fraction correct between cohorts. For one cohort, a statistically significant increase with large effect size in fraction correct between cohorts was found, which supports the hypothesis. Results from a recently published paper in Advances in Engineering Education will be enhanced by data from more recent cohorts in this contribution.

108d Python for Chemical Engineering Calculations: An Elective Course for Undergraduate Students

Gennady Gor

Modern engineering calculations require a flexible and efficient programming language. Python is such a language: it is open source, free, easy to learn, and simple to use. These factors make Python one of the most popular programming languages in the world, highly demanded by employers. However, most undergraduate programming courses for engineers focus on languages other than Python.

I developed and taught a new course 'Python for Chemical Engineering Calculations', which was offered in Spring 2021 as a 3-credit undergraduate elective.[1] The goal of this course was to introduce undergraduate chemical engineering students to Python and demonstrate how it can be used for solving a spectrum of chemical engineering problems. The example problems were taken from the undergraduate chemical engineering curriculum, e.g., from courses such as Chemical Engineering Thermodynamics, Fluid Flow, Kinetics and Reactor Design, etc. Lectures and practical sessions were complemented by six guest lectures delivered by engineers working in industry who demonstrated the use of Python in their jobs. I used the 'flipped classroom' approach, where the students watched the short tutorial videos before each class.[2] Classes typically started from short quizzes based on the videos, after which the class time was utilized for hands-on activities.

The course was very well received by the students. They spoke highly of the strong connection between programming and the chemical engineering curriculum, which was impossible to see from taking a generic programming course. Guest lectures was another aspect enjoyed by many. Finally, the flipped classroom which provided a lot of time for in-class activities, appealed to the students a lot. This success suggests that some of this course materials and approaches can be utilized for teaching the required programming and numerical methods courses.

1. G. Y. Gor, ChE 490: Special Topic - Python Programming for Chemical Engineers. Chemical and Materials Engineering Syllabi. https://digitalcommons.njit.edu/cme-syllabi/160
2. G. Y. Gor, ChE490: Python for Chemical Engineering Calculations. Video Lectures. https://www.youtube.com/playlist?list=PLWx-kn1Xf12a77HKLCtZn4E2cjjjoL8N8

108e Experience with Symbolic Algebra and Smart Computing in Chemical Engineering Curriculum

Satish Parulekar

Experience in using symbolic algebra and smart numerical computations in undergraduate and graduate chemical engineering courses using Mathcad and Matlab is discussed to improve understanding of process fundamentals. Emphasis is on obtaining solutions with increased confidence and higher precision and with minimum effort by understanding what type of problem is being presented. Students are encouraged to not introduce unnecessary complications when attempting to solve a problem and solve problems using optimal effort. Example problems considered for illustration deal with solution of linear algebraic equations (kinetic parameter estimation and properties of small and large reaction schemes and metabolic networks), nonlinear algebraic equations (vapor-liquid equilibria for ideal and non-ideal systems, equilibrium calculations for multiple reactions, isothermal and non-isothermal operations of well-mixed reactors with single and multiple reactions, and kinetic parameter estimation using nonlinear regression), ordinary differential equations (isothermal, adiabatic, and non-adiabatic batch and steady state continuous flow tubular reactors), application of difference equations, and symbolic algebra (rate expressions for catalytic reactions and characteristics of reaction networks). Analysis of multi-reaction networks is based on properties of atomic and stoichiometric matrices. A smart use of relations among atomic and molecular species and molecular species and reactions obviates the need for solving conservation equations for all species in the reaction network, which typically are systems of nonlinear algebraic equations and nonlinear ordinary and partial differential equations. Any independent reactions missing in the reaction network are identified in a systematic fashion. Reduced order reactor models for large reaction networks can be obtained by applying pseudo-steady state hypothesis for highly reactive species, such as free radicals and various forms of catalytically active sites, an approach also useful in flux balance analysis in networks of reactions in living cell cultures. Symbolic algebra is very useful in this effort. Mathcad is interactive, with user-friendly graphics capabilities. Students are encouraged to display results using graphics for improved learning and understanding. The capabilities of Mathcad are of significant benefit in accelerating the learning and strengthening the fundamental knowledge base.

108f Open Source Python-Based Application As an Interactive Spectroscopic Teaching Aid

Jakub Konkol, George Tsilomelekis

Spectroscopic techniques are unquestionably valuable for understanding structural and compositional changes in complex chemical, biochemical and biological systems. Consequently, spectroscopic techniques are increasingly incorporated to undergraduate/graduate curricula to prepare the next generation of engineers for successful careers in industry and academia where applications include reaction monitoring, quality control, and process control etc. Spectroscopic pedagogy excels in theory and interpretation of spectra and some educators are shifting towards programming-based lessons to introduce certain topics. The interactivity and exposure of computer science in the field of chemistry are advantageous in a future scientist's career.

Among the challenges in traditional spectroscopic pedagogy, two stand out: the practical aspect of collecting meaningful and reproducible data along with the relevant and necessary pre-processing prior to any interpretation. Lab-based sessions are expensive and certain techniques may be prohibitively long or sensitive for students to broach topics such as instrumental parameters or background drifts. Through programming-based modules developed for education, these topics can be introduced earlier to make higher education level students aware of the reality of spectroscopic techniques in addition to the theory. Continuing our previous work, we have transitioned our previous application to Python to better utilize an open-sourced framework. In addition, the application is further enhanced in this transition for the educator by incorporating chemometric analysis, data extraction, as well as advanced preprocessing techniques that will assist in teaching Raman spectroscopy, such as the Stokes – Anti-Stokes temperature ratio. The benefits and pitfalls of developing using Python, in the lens of previously using MATLAB, are also discussed.

108g Hello World of (Bio)Chemical Engineers: A Structured and Tailored Python Course

Fiammetta Caccavale, Carina L. Gargalo, Krist V. Gernaey, Ulrich Krühne

The Fourth Industrial Revolution refers to the transformation of industry through automation and digitalization. In order to meet the requirements imposed by this new digital era, engineers need to be adequately prepared and trained [1, 2]. Therefore, there is an emergent need to introduce new technologies in the educational system. Education, at the same time, has also faced a wave of digital transformation due to the COVID-19 crisis [3]. This forced educational institutions worldwide to adapt and switch to E-learning quickly.

Previous studies [4] highlight the need expressed by students to acquire programming skills in (bio)chemical engineering education. This quantitative survey sheds light on different issues, such as the insufficient presence of programming content in the curriculum. There is a clear urgency to introduce and build on programming knowledge, especially targeting topics such as modelling and optimization, which are, and have always been, in high demand. Moreover, the study presented in [4] also points out that the students prefer to use only one programming language - Python being the favored one over MATLAB.

Other aspects that make Python a better choice for students and engineers that want to adapt to this digital transformation is that this language is one of the most popular languages in use today, thanks to its straightforward syntax, readability, available libraries, broad documentation, support, and suitability for data science applications. Even in the chemical engineering community, Python is attracting the interest of both students and professionals. Furthermore, it is regarded as the to-go language for data analysis and data science, both in academia and industry. We confirmed this by looking at the repositories available in GitHub obtained through a keyword search, including 'chemical engineering'. Among the results, 38.3% of repositories use Jupyter Notebooks (Python), 20.3% pure Python, and only 9.4% MATLAB, which historically has been the preferred language used in chemical engineering.

Although online resources to learn Python are abundant, tailored courses for chemical engineers tend to scratch the surface of syntax. At the same time, courses for a more broad audience might be too general for someone that wants to learn operations, modelling and discover available packages to solve chemical engineering tasks.

To tackle and address the mentioned challenges, in this work, we propose a hands-on Python programming course tailored to (bio)chemical engineers. The course's objectives are to: (a) teach the basics of Python programming; (b) impart more in-depth knowledge by providing real use cases and applicable/practical tools; and, (c) inform and inspire about possibilities given by different tools. Furthermore, this course is being created based upon the following three main pillars: (i) relevance and usefulness, with focus on problems within the chemical engineering field; (ii) informativeness, aiming to provide thorough explanations, background, fun facts, links to bonus material; and, (iii) entertainment and practicality, through stimulant exercises and gamification, students should learn by programming, embedded reward through difficulty levels.

In order to evaluate the usefulness, applicability, and relevance of the course, a preliminary schedule and objectives were presented to the faculty and staff of the PROSYS research centre at the Department of Chemical and Biochemical Engineering at the Technical University of Denmark (DTU). This test group was invited to give feedback on the overall course and on the topics within. Additionally, the test group was invited to fill in a survey suggesting the topics believed to be relevant for the course. The results have been synthesized and are presented in the word cloud in Figure 1. This co-participatory experience allowed us to gather many valuable requests for material to be incorporated in the course.

Hence, among other topics, the plan is to incorporate how to use Python to, for example, solve mass and energy balances, differential equations, and optimization problems. Additionally, the course will contain comprehensive explanations and examples of: (i) data processing, (ii) data analysis, (iii) data science, and (iv) multiple use cases of how machine learning, natural language processing, and image processing can be applied in (bio)chemical engineering.

References

[1] C. L. Gargalo, S. C. d. l. Heras, M. N. Jones, I. Udugama, S. S. Mansouri, U. Krühne, and K. V. Gernaey, "Towards the development of digital twins for the bio-manufacturing industry," Digital Twins, pp. 1–34, 2020.

[2] H. Narayanan, M. F. Luna, M. von Stosch, M. N. Cruz Bournazou, G. Polotti, M. Morbidelli, A. Butté, and M. Sokolov, "Bioprocessing in the digital age: the role of process models," Biotechnology journal, vol. 15, no. 1, p. 1900172, 2020.

[3] UNESCO, 'The impact of the COVID-19 pandemic on education: international evidence from the Responses to Educational Disruption Survey (REDS)' 2022.

[4] S. C. de Las Heras, C. L. Gargalo, K. V. Gernaey, and U. Krühne, "Programming skills across the (bio) engineering curriculum–a students' perspective," in Computer Aided Chemical Engineering, vol. 50, pp. 2039–2044, Elsevier, 2021

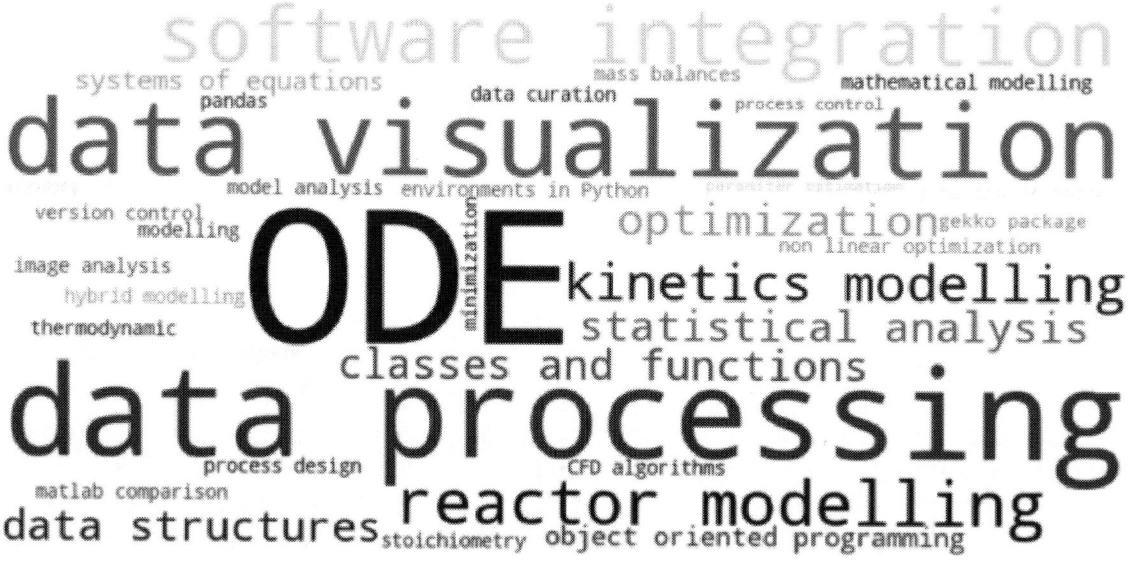

Figure 1: Word Cloud presenting the topics suggested to be included in the Python course by the faculty and staff of the PROSYS research centre at the Department of Chemical and Biochemical Engineering at the Technical University of Denmark (DTU).

49a How We Incorporate the Impact of Engineering Solutions in Global, Economic, Environmental and Social Contexts

Taryn Bayles, Joaquin Rodriguez Alonso, Robert M. Enick

In order to meet ABET student outcomes (2) "the ability to apply engineering design to produce solutions that meet specified needs with consideration of public health, safety and welfare, as well as global, cultural, social, environmental and economic factors" and (4) "ability to recognize ethical and professional responsibilities in engineering situations and make informed judgements, which must consider the impact of engineering solutions in global, economic, environmental and social contexts" our department has incorporated various assignments and projects into our required courses so that our students the opportunity to demonstrate that they have achieved these outcomes.

In this presentation, we will provide examples of homework assignments, quizzes, and projects (with grading rubrics) which have been added to our sophomore and junior level core classes. These assignments provide students with the foundation they need to complete larger scale projects during their senior year, in three capstone courses (Process Control, Safety andamp; Ethics, and Process Design). In the Process Control course, students complete a Global Project in which they collaborate with a foreign organization to become acquainted with a problem in that foreign country. The students then analyze the problem and propose a solution for the challenge. Students in the Safety andamp; Ethics course are required to attend a three-hour class on the Deepwater Horizon incident (which is given by one of our faculty members who served on the flow rate estimation team and also surveyed a multitude of reports on this incident). Each student team must then report on the technical, ethical, economic, global, environmental, and regulatory aspects of the Deepwater Horizon incident and the related efforts to stop the leak, clean up the spill, and prevent similar occurrences. In addition, each team must attend a three-hour lecture related to a storage tank explosion and then describe the technical ethical, social, cultural, global, economic, and environmental impacts of such an incident as they design simple strategies to prevent a re-occurrence of this fatal incident. And finally in the Process Design class, students are required to complete a design project over the course of the semester, which includes process synthesis, equipment design, process optimization andamp; control, and economic analysis of their chemical plant. Students propose a location for their plant, and are also assigned a second location in a different continent, and compare and contrast the impact of their design in a global, economic, environmental, and social context for the two different locations. Details of these four projects will be shared, as well as examples of student work.

49b Incorporating the Stories of Scientific Discovery into the Chemical Engineering Curriculum

Stephanie Velegol

The early 20th century brought incredible discoveries such as the production of ammonia, nuclear fission and the discovery of polymers. Although students often learn about these concepts, these don't always hear about the societal impact or the stories of the people behind the discoveries. For example, Fritz Haber engineered a way to create ammonia from nitrogen and hydrogen and then used his skills with high pressure gas to release chlorine in the battle of Ypres in 1915. His use of chemical warfare caused his wife to take her life. Otto Hahn and Lise Meitner discovered fission with others in 1938 and yet only Hahn had his name on the paper, and he alone won the Noble Prize due to societal pressures. Decades later, element 109 was named after her. Hahn was one of ten captured scientists in Farm Hall in England when the atomic bomb was dropped on Hiroshima. Recordings show that he was distraught that his discovery brought so much destruction. Polyethylene was discovered by accident in both 1898 and 1933 and has been used for beneficially for society since then. At the same time, there is now an increase in plastic pollution.

The purpose of this talk is to share how these stories can be used throughout the chemical engineering curriculum to help students think about the impact of society on science, the impact of science on society and the human stories behind each discovery. Helpful resources will shared, such as articles, websites, podcasts, docudramas and books.

49c Improving Scientific and Risk Public Communication of Process Safety Incidents for Chemical Engineering Undergraduates

Amanda Koh, Lance Simpson

Today's climate of scientific skepticism and confusion makes it clear that the ability of scientists and engineers to clearly, simply, and effectively communicate science, risk, and safety is not only incredibly important but generally underdeveloped. A likely avenue for this type of skill building would during an engineer's undergraduate education. With a similar eye to the needs of modern chemical engineering, and in accordance with U.S. Chemical Safety Board recommendations over the past decade, most chemical engineering undergraduate programs incorporate some amount of process health and safety instruction into their curriculum. While the incorporation of process safety into the chemical engineering curriculum has been valuable, the standard set of topics only touch on how to understand the impacts of chemical process incidents to the community which invariably interacts with and surrounds the plant (i.e. voluntary vs. involuntary risk, dispersion modeling of toxic material into the community, etc.). Chemical process health and safety, however, is a uniquely advantageous class to teach students how to communicate technical information that is immediately relevant to the daily life of a "layperson." In this work, we develop a class module that takes advantage of expertise in other departments such as communication, risk management, and science literacy to teach undergraduate chemical engineers to discuss a prior (real-life) chemical process incident with a mock town hall. The town hall is comprised of both those with and without domain-specific expertise, and the town hall participants are asked to act as members of the effected community. The students are assessed on their ability to clearly communicate the basic information of the incident and the strategies moving forward to improve safety as well as their ability to be perceived as competent, confident, and able to answer concerned resident questions. The results of this work will be used to expand the course module to larger groups of students as well as inform better, long-lasting strategies to teach engineers and scientists to better communicate science, risk, and safety to the non-technical public. If successful, this type of skill building will have enormous positive impacts on the type of public discourse society has related to everything from vaccines to climate change to genetics.

49d Incorporating an Industry Role-Playing Experience in Laboratory Classes to Improve the Technical Communications of Students

Stephanie Wettstein, Jennifer Brown, Michelle Miley

Professional skills, including problem-solving, project management, team management, and communication, are highly valued in industry and yet difficult to incorporate effectively into the curriculum. In particular, written and oral communications skills are lacking, which leads to less career advancement opportunities for students that have lower abilities. Starting in 2017, changes were made to a senior-level laboratory course to mimic an internship experience in order to motivate students to become better technical communicators. Changes included making lectures "workshops." incorporating the University's Writing Center, creating memorandum-style homework, and making the lab objectives industrially relevant. These relatively small changes to the course made big impacts to the students' writing, and student feedback has been overwhelmingly positive. Instructors from the second lab course noticed a large difference in the quality of writing due to the changes made and students have emailed with compliments they have received on their communications after graduation. This work was awarded the 2022 Innovation in Teaching award at MSU due to the transformation of this senior lab course and an NSF IUSE grant was awarded to study the effects of further incorporating industry aspects into laboratory course. This presentation will outline the recent improvements in the senior lab I course at MSU and how it helps motivate students to improve their technical communications.

49e Initial Offerings of an Undergraduate Research Methods Class: Results and Experiences

Joseph Holles

In an effort to improve university recruiting and retention, our institution has focused on expanding the Honor's College experience. This includes individual Honors experiences in each college and the University's Honors College. The goal within the College of Engineering and Applied Sciences (CEAS), is to focus the Honors experience on undergraduate research with an aim of broadening research opportunities and competitiveness of student applications for summer research programs, NSF REUs, internal/external research funding applications, participation in undergraduate research conferences, and preparing the students for graduate school. Many students (inside and outside of the honors program) have received credit for completing undergraduate research, but this is often a "stand-alone" course with no additional preparation and ill-defined outcomes. This approach does provide an opportunity for a laboratory experience, the research experience is greatly dependent on the mentoring provided informally by the research laboratory and the advisor. Parallel to the CEAS focus, the Honors College would like to increase the number of students completing their Honors Capstone (senior thesis/project). To respond to these goals, an "Introduction to STEM Research" course was developed and taught. This course was broadly designed to group mentor the students into the research process, find an advisor, prepare them for the subsequent in-laboratory research experience, and scaffold them towards completion of the Honors program. This course is broadly focused to provide a general approach to research and graduate school preparation appropriate for all majors in the Engineering College and other programs in hard science.

The Research Methods course has now been taught twice: Spring 2020 and Fall 2021. While offered through the Honors College/Engineering Honors, the course was open to all interested students. For the initial two offerings, there have been a total of nine students (six engineering and three others). Course topics included: finding a research mentor, literature search skills, using the scientific method for approaching a research problem, developing a research methodology, writing a funding proposal, delivering a research presentation, and selecting and applying for graduate school. Results, experiences, observations and student feedback from these initial offerings will be presented. All students found one (or more) research advisors and initiated their research in positions from volunteer, to for credit, and paid researcher. All students completed the required university safety training. Several students submitted successful applications for funding or undergraduate research programs. Student self-reported levels of experience on course elements were collected pre- and post-course. The average response increased by 1.31 points, and 22 pre-course "Don't Knows" became zero "Don't Knows" post-course. Results from the CURE tool for examining classroom aspects of undergraduate research were also used as a tool to benchmark this course against similar approaches to exposing undergraduates to research. "Benefits" from the CURE survey were compared to this course. Results indicated that this course helped prepare the students for undergraduate research similar to other courses and programs already included in CURE with similar goals. As such, this course offers an additional tool for educators to prepare students for research.

49f Professional Development for Students through Company Outreach

Jean Tom, Jennifer Lott, Mara Giga, Masano Huang

Evaluating and understanding available opportunities in chemical engineering are important components of professional development for undergraduate students. To support this type of development, a group of women chemical engineers at Bristol Myers Squibb (BMS) has developed a grassroots outreach program to expose students to chemical engineering careers in the pharmaceutical industry. The event, held annually at Bristol Myers Squibb (BMS) for the past 9 years, has generated survey data from the students. The data has yielded insights on the program's impact to (1) increase an understanding of what chemical engineers do in the pharmaceutical industry, (2) present a broader view of academic options, (3) build higher confidence in the students' choice of major, (4) understand the impact of imposter phenomenon, and (5) foster a sense of belonging and pride to be a woman in a STEM field. The in-person (first 7 years) and virtual (last 2 years) event includes seminars, poster sessions, lab tours, networking sessions, and discussion panels. We present data from ~250 women chemical engineering students who participated in the events. While these events enhanced the company's ability to attract and hire top talent and engaged current employees as role models for the next generation of women in STEM, they can also supplement the formal chemical engineering education to expose students to the professional skills needed in the workplace and how to frame decisions about their career choices.

49g Coacheng: Virtual Mentoring for Chemical Engineering Students

Bryan Alamani, Jhud Mikhail Aberilla, Rowena Ordonez-Torres, Marjorie Baynosa

Remote learning has posed many challenges into the delivery of quality education for chemical engineering students. However, this remote setup has also opened up opportunities to rethink ways on how learning can be made holistic and cross borders and time zones. Here, we are presenting our CoaChEng program, an E-mentoring program at the University of the Philippines Diliman Chemical Engineering program. CoaChEng is a joint effort of alumni and faculty of the Department aimed at providing relevant one-on-one mentoring sessions for undergraduate students taking up chemical engineering. CoaChEng delivers dedicated sessions on topics such as communication skills, advocating, collaboration, among others. The goal of the program is to provide complementary non-technical skills to undergraduate students using remote learning platform and online tools. Insights on the challenges and opportunities from e-mentoring shall also be discussed.

49h Study of Upper Division Chemical Engineering Students' Motivations in Selecting a First Position upon Graduation

Victoria Kolar, Allison Godwin

Engineering education often focuses on developing students for engineering careers, and much of the current body of literature focuses on understanding factors for preparation to become professional engineers. Most of these studies focus on mastering knowledge, skills, and abilities that educators and employers think are important [1]. Fewer studies approach the problem by exploring how young engineers perceive the relevance of their education and the process of a job search. Often, these studies have focused on early-career engineers in the workplace, while a few have studied engineering students currently in school. One study, the Professional Engineering Pathways Study (PEPS), examined junior and senior students at six institutions through their graduation and into their first year using a survey and follow-up interviews [2]. This study found that students' beliefs about their ability to succeed in engineering and their value of engineering tasks had a significant influence on career selection. Fewer studies have investigated how students conceptualize their future careers and how they make decisions about which industry sectors and to which companies they apply. In addition, these factors are likely different by engineering discipline–student pathways post-graduation and disciplinary norms are varied [3]. Additionally, studies in engineering education research, broadly, and in chemical engineering education, specifically, indicate that students' beliefs about ability to succeed and career values may differ by gender [3-5]. Often, women have lower expectancy beliefs than their male peers and different career values. These differences are often attributed to both gender socialization as well as valued norms in engineering [5]. Together, this prior work points to a need to better understand how students consider future careers and which factors they consider in the job search, particularly in a chemical engineering context.

This study investigated upper-level chemical engineering students' beliefs about their abilities to succeed in engineering careers and their values of particular factors in choosing specific engineering careers in their first entry level job through the following research questions:

1. Which factors are most important for upper-level chemical engineering students in selecting their first position after graduation?
2. How does particular company values, expectancies, and prior work experience influence these factors? And, do these factors differ by gender?

We used the framework of expectancy-value theory to understand how students may be motivated to set particular goals in a job search and how beliefs about ability to succeed (i.e., expectancy) and personal values may affect actions taken towards goal setting and the job application process. Expectancy-value is a motivation theory that frames motivation for a given behavior, action, or goal is defined by expectancy (i.e., beliefs about ability to succeed on the task) and value (i.e., how much the individual values the desired outcome) [6].

The data for this study come from a survey of 85 junior and senior chemical engineering students at a large, public Midwestern university. Students were asked to complete an online survey without incentives for participation (the study was approved by the IRB). Items measured their expectancy on chemical engineering tasks both in and outside of the classroom on a five-point Likert scale. In addition, students were asked to rate a list of factors that were important to them in a future job including: Mentorship/Role Models at the Company, Compensation, Company Culture, Location, Company Benefits, Job Expectations, Promotion Opportunities, Future Higher Education and

Professional Development Opportunities, Travel Opportunities, and Other (write-in if selected). Students were also asked to rate the company values that were important when considering a company in an application process, which included: Integrity, Commitment to Diversity/Equality, Respect, Loyalty, Honesty, Trust, Collaboration, Service in the Community, Safety, Other (write-in if selected). The rating lists for both personal values and company values were generated from a literature review, examination of company websites, and pilot qualitative discussions with students. Students were also asked to self-report career interests; interest in job sectors in chemical engineering; co-op or internship experiences; and demographics. Demographic measures included non-binary gender and multi-select race/ethnicity options.

Multiple linear regression models were used to examine how students' expectancy beliefs and value of particular company values shaped their rankings of the most important factors in a future job. Interaction effects by expectancy and company values as well as gender and expectancy were also included in these analyses based on prior literature emphasizing that women, in particular, may have different career goals and lower expectancy beliefs in engineering contexts [3]. A stepwise modeling approach was used, where all hypothesized factors were added into the model and backwards deletion was used to remove non-significant results (alpha value set at 0.05). We first modeled the main effects of company values, gender, expectancy, and participation in an internship or co-operative education experience. Then, we modeled the theorized interaction effects for gender and expectancy.

The results indicated that the most highly rated factor for chemical engineers when considering their entry level job was perceived company culture (mean rank = 3.133). The company values of Service in the Community (B =.366, $p < .001$), Trust (B =.308, p =.012), Respect (B =.309, p =.012), Commitment to Diversity/Equity (B =.197, p =.024), and Safety (B =.211, p =.025) significantly predicted higher rankings of company culture. Students' confidence in their abilities to secure their dream job after graduation was also a significant negative predictor of ranking company culture highly (B =-.454, p =.046). Additionally, students who had prior co-op or internship experiences were significantly more likely to rank Compensation as an important consideration in their careers (B =1.371, p =.009). We found several interaction effects between students' expectancy beliefs in chemical engineering work both for undergraduate education and securing a future job and company values for predicting the most important factors for consideration in future careers, particularly in relation to Respect as a company value for predicting the importance of company culture in students job seeking.

Additionally, the only significant interaction effect by gender and expectancy was in predicting the ranking of promotion opportunities highly in consideration of a job search. Men who indicated a higher confidence in their ability to understand their chemical engineering core classes were less likely to rank promotion opportunities as important in their list of job search priorities (B=-.738, p =.0140). In contrast, women who indicated a high expectancy in becoming a successful chemical engineer were more likely to rank promotion opportunities as important in their list of job search priorities (B=.612, p =.0195). These results indicate how one job factor, promotion opportunity, had different priorities by both gender and expectancies in chemical engineering curricular contexts versus a larger set of career successes.

Together, these results emphasize how particular company values may shape students' perceptions of and motivation to pursue particular jobs in their application processes. Additionally, the may highlight particular gaps between the current graduating cohorts from chemical engineering programs and existing companies. These conclusions can be used by companies to attract the best possible chemical engineering candidate that provides a fit for both the company and employee. These results also provide information on a transition that is less often studied in chemical engineering education.

Literature Cited

[1] Gilmartin, S. K., Antonio, A. L., Brunhaver, S. R., Chen, H. L., andamp; Sheppard, S. D., "Career Plans of Undergraduate Engineering Students: Characteristics and Contexts," Freeman, R. B. andamp; Salzman, H., eds., "U.S. Engineering in a Global Economy," University of Chicago Press. (2018).

[2] Brunhaver, S., Streveler, R., Carrico, C., Matusovich, H., Boylan-Ashraf, P., andamp; Sheppard, S., "Professional engineering pathways study: A longitudinal study of early career preparedness and decision-making," Proceedings of the IEEE Frontiers in Education Conference, El Paso, TX (Oct. 21-24, 2015).

[3] Verdín, D., Godwin, A., Kirn, A., Benson, L., andamp; Potvin, G., "Engineering women's attitudes and goals in choosing disciplines with above and below average female representation," *Social Sciences*, **7**(3), pp. 44-69 (2018).

[4] Brawner, C. E., Lord, S. M., Layton, R. A., Ohland, M. W., andamp; Long, R. A., "Factors affecting women's persistence in chemical engineering," *International Journal of Engineering Education*, **31**(6A), pp. 1431-1447 (2015).

[5] Godwin, A., Verdín, D., Kirn, A., andamp; Satterfield, D., "The intersection of gender and race: Exploring chemical engineering students' attitudes," *Chemical Engineering Education*, **52**(2), pp. 89-97 (2018).

[6] Brown, P. R., McCord, R. E., Matusovich, H. M., andamp; Kajfez, R. L., "The use of motivation theory in engineering education research: A systematic review of literature," *European Journal of Engineering Education*, **40**(2), pp. 186-205 (2015).

239a How the ChE Education Community Is Satisfying the Institute's Program Criteria Associated with Design, Analysis and Control of Processes: Survey Data from Program Evaluators

Donald Visco Jr., Thomas Spicer III, Troy Vogel

In this presentation, the authors analyze anonymous ABET program evaluator feedback considering trends and best practices (in an anonymized fashion) of how chemical engineering programs have attempted to satisfy the AIChE Program Criteria related to "*engineering applications ..to the design, analysis and control of processes, including the hazards associated with these processes*". Such an aggregation will present perspective to the individual program presentations as part of this session.

239b Integrating Chemical Process Safety into a Chemical Engineering Curriculum

Richard Davis, Weiguo Xie, Michael Rother, Tsutomu Shimotori

In response to industry needs, ABET and our lead society AIChE set program accreditation criteria for the curriculum to include the design, analysis, and control of processes, including the hazards associated with these processes. [i] The impetus for this criterion stems from catastrophic accidents in the chemical process facilities that brought chemical process safety (CPS) to the national forefront. These accidents were attributed to the lack of training of engineers involved in their design and operation.[ii] The chemical industry and government agencies for oversight and investigation of incidents have recommended increased emphasis and training of chemical engineers in CPS. The result of these recommendations is the inclusion of safety in the criteria for the accreditation of Chemical Engineering programs by ABET.

The department faculty is committed to teaching chemical process safety. To sustain the program, we recognize that we are accountable to our constituents, including our faculty and staff, students, and industries that employ our graduates. We have enlisted the assistance of process safety professionals from the industry to provide guidance and feedback on our program enhancement activities. This integrated approach is sustained by our annual accreditation activities and reporting to and feedback from our department's IAB.

Our approach to meeting these criteria integrates topics of CPS into the core curriculum of our four-year BSChE degree and graduate programs. By integrating safety into all program levels, we elevate the importance of CPS to the level of other pillars of the discipline, including chemistry, physics, and mathematics. We expect our approach fosters the skill of lifelong learning and commitment to CPS in our graduates.

Our primary objectives are for chemical engineers to graduate with a background in CPS and a safety culture. Engineers with a safety culture naturally act safely and encourage others to act safely. Joe Louvar wrote an editorial for the Dec 2012 issue of Process Safety Progress. He warned, "...our university students really don't understand process hazards analysis or concepts of inherent safety. This is especially disappointing because these concepts are of paramount importance in an industrial environment."

Many programs at other universities have responded by adding a course on chemical process safety into their curriculum. Our challenge with this approach is deciding what to eliminate from an already crowded curriculum. After several consultations with our department's Industrial Advisory Board (IAB), we have determined to take a different approach by integrating safety instruction throughout our curriculum to emphasize the far-reaching importance of safety in practice and instill a safety culture in the department that stays with our students after graduation.

We have identified the following safety-related student learning outcomes. Our graduates:

1. Understand the importance of CPS and required resources/commitment
2. Can characterize hazards associated with chemicals/agents
3. Understand/apply concepts of an inherently safer design
4. Understand how to control/mitigate hazards to prevent accidents
5. Are familiar with major regulations

6. Understand consequences of incidents due to acute/chronic releases/exposures
7. Are proficient with hazard identification procedures
8. Have an introduction to the process of hazard evaluation/risk assessment

Another specific outcome for senior design is a systematic approach to assessing and discussing safety in the team presentations and reports. Students will have incorporated inherently safe design at the end of the capstone design course as an integral part and underlying theme of their projects and engineering work.

To reach these goals, we:

- Provide our instructors and students with access to digital media for safety training through SAChE modules. The SAChE modules are assigned to match the core program courses required by all of our majors.
- Work with our instructional staff to design and implement CPS active learning in their courses. These include:
    - student-produced safety training videos
    - student enrollment in AIChE to complete SAChE's online safety certificate program
    - homework problem sets and projects around CPS
    - students' problem-based learning in teams to submit a literature review on CPS and present their findings to the whole class
    - a capstone design project that requires consideration of safety constraints and design for inherent safety
- Employ practical assessment methods, including rubrics for student outcomes, specific CPS performance indicators, and qualitative surveys.

We integrate teaching and learning of CPS across our curriculum by leveraging digital media for safety training available from a variety of resources, including the University of Minnesota Libraries and offices of Environmental Health and Safety, and with pedagogy for digitally enhanced instruction in partnership with the University of Minnesota's Center for Teaching and Learning. We also draw from digital media for industrial chemical process safety training available from the US Chemical Safety Board, Occupational Safety and Health Administration, American Institute for Chemical Engineers' Center for Chemical Process Safety (CCPS), Environmental Protection Agency, and industry-produced digital resources.

Our assessment plans match our campus assessment reporting requirements to streamline data-driven program assessment and continual improvement.

[i] https://www.abet.org/accreditation/accreditation-criteria/criteria-for-accrediting-engineering-programs-2022-2023/#3

[ii] Chemical Safety Board, *Investigation Report, T2 Laboratories, Inc., Runaway Reaction*, Report No. 2008-3-I-FL, Sept., 2009

239c Chemical Process Safety: A 3-Credit Course at the University of Florida

Mark E. Orazem, Vincent Tocco Jr., Helena Hagelin Weaver

Our department has a three-credit core course on chemical process safety. It has evolved from a one-credit course associated with our unit operations laboratory to a required two-credit course (Spring 2010), *Safety and Experimental Evaluation*. After the inclusion of safety into the ABET Student Outcomes, three major changes were made, 1) The course name was changed to *Chemical Process Safety* to better describe the course content, 2) the number of credits were increased from two to three, and 3) we made it a core course, which requires a minimum grade of C for passing. Each of us has served as lead instructor for the course, and, while each of us has our own style and emphasis, the common features are:

1. We use Crowl and Louvar (Chemical Process Safety: Fundamentals with Applications) as our textbook. Students are free to use the third or fourth editions, but not earlier editions.
2. We make ample use of the resources provided by the Chemical Safety Board (CSB). This is done by showing CSB videos and discussing CSB investigations in class.
3. The Chemical Reactivity Worksheet program developed by NOAA is introduced, and assignments are included where students use the program to identify incompatible chemicals.
4. SAChE certification is motivated by awarding up to a letter grade of extra credit (e.g., 5+ certifications = 1 letter grade). As the average grade for the class before extra credit is given is around a B, this means that most students will receive an A in the class. We require students to take the *Chemical Reactivity Hazards* and *Process Safety Ethics – A Brief Introduction* courses, while other certifications for extra credit may be selected among the level 2 and 3 courses.
5. We use homework and exams to motivate students, and often design exam questions around recent accidents investigated by the CSB.

Features included by some instructors include:

1. Guest speakers, including a speaker from OSHA who has prior experience working with the Chemical Safety Board, an environmental health and safety expert from Bacardi, and a retired vice president from BP. Their presentations help to emphasize the real-world challenges of managing safety.
2. Introduction to the NOAA programs (ALOHA, MARPLOT, CAMEO).
3. Assignment of a written essay of at least 400 words, with the prompt "Describe the ethical and professional responsibilities of Chemical Engineers, and the impact of engineering solutions in environmental or societal contexts."

This course is taught after students have taken material and energy balances, thermodynamics, and transport phenomena. Chemical process safety is a mid-level capstone course as students see how the material they have learned to date can be used to recognize and mitigate safety hazards in industry. Our approach to teaching safety is both traditional and effective.

Key Words Education, Process Safety Competency, Process Safety Culture

A Multi-Dimensional Approach: Applying Process Safety Principles in Several Courses Across the Curriculum

Stephanie D. Loveland, R. Dennis Vigil, and Derrick K. Rollins
Department of Chemical and Biological Engineering
Iowa State University, Ames, Iowa

Introduction:

The subject of process safety instruction has gained attention since the introduction of the ABET/AIChE requirements to include hazard awareness in undergraduate chemical engineering curricula in 2012 [1]. Chemical engineering departments have addressed this requirement in several different ways. Some have incorporated safety instruction into the existing core curriculum, some have developed dedicated process safety courses, and others have applied a combination of the two [2]. The combination method has been suggested to be the best option as it follows best practices in integrated learning methods and allows reinforcement of concepts in multiple courses rather than compartmentalizing concepts into separate modules [3]. This paper describes the approach used in the Department of Chemical and Biological Engineering at Iowa State University and our future plans to continue to develop higher levels of process safety instruction to produce competent chemical engineering graduates.

Overview of Process Safety instruction at ISU:

In 2009, the CBE department at Iowa State University began to incorporate some safety modules into specific courses in the curriculum. These included the discussion of runaway reactions in the reactor design course, proper valve and controller selection in process control, and general awareness of materials selection and safety-related design principles in process design. With the implementation of the 2012 ABET requirement to incorporate hazard awareness into the undergraduate chemical engineering curriculum [1], these efforts became more focused and organized. A dedicated course offering was introduced as an elective, and this eventually became a graduation requirement. But some concepts are still incorporated into other courses in the curriculum, in order to help reinforce the ideas and emphasize the importance of safety for chemical engineers.

The courses that have been selected to incorporate safety modules or lectures were chosen because of their place in the program of study (when students take them) and subject matter. Students will be exposed to process safety concepts starting in the second-year seminar course, and then several times throughout the rest of their studies, ending with the process design course and final unit operations laboratory course, which are taken the final semester of the program of study. While effort is made to highlight process safety concepts throughout the four-year curriculum, most students will see the most process safety content during their senior year.

First Exposure: ChE 202

ChE 202 is a seminar course for sophomore students that introduces them to many different professional requirements. It includes discussions on ethics, an overview of different career opportunities for chemical engineers, and other life skills. The purpose of the Process Safety lecture in this course is to provide historical context of major process safety incidents so that students can understand how these events have affected the practice of chemical engineering, and their curriculum at Iowa State. In this lecture, students learn about early efforts at DuPont (1912) to begin implementing a safety program [4]. The Flixborough (UK, 1974) and Seveso (Italy, 1976) incidents are discussed to show how public awareness of major incidents began to shape our profession. The Bhopal (India, 1984) and Piper Alpha (UK, 1988) incidents are discussed in a little more detail, and students learn about their influence on US regulations that were created following these events. Finally, the BP Texas City (Texas, 2005) and T2 Laboratories (Florida, 2007) incidents are presented, and the resulting changes to the ABET criteria (2011) that have influenced their coursework are noted.

ChE 325 and 426: the Unit Operations Laboratories

The next place most students are faced with process safety concepts is in the Unit Operations laboratory course ChE 325. This is taken during the junior year, after the first two transport courses and the chemistry and chemical engineering thermodynamics courses. The laboratory is a typical place that many programs incorporate safety concepts [3], but it's important to make sure that the differences between lab/personal safety and process safety are highlighted. In the lab, when students are conducting regular experimental work, there are a number of safety items they are required to address. First are the general personal/lab safety rules and culture that require proper PPE and attire, housekeeping, awareness of chemical storage guidelines, and handling of compressed gases. Students take three EHS training modules to address these things. The next requirement is that students are aware of the process safety hazards that exist with the different experiments that they will operate in the laboratory. In order to keep track of these hazards, each student must complete a Job Safety Assessment card whenever they are working in the laboratory. These cards ask specific questions about chemical, physical and electrical hazards associated with the equipment for each experiment. The intent of using the cards each time students are in the laboratory is to encourage awareness of these hazards and remind students of the mitigation steps necessary to help prevent an incident and protect them should one occur.

At the end of the semester during the junior lab course, students design an experiment and present a proposal to the laboratory staff. Part of this proposal requires students to consider the safety of the experiment, apply inherently safer design principles, and implement any necessary safeguards (preventive and mitigative) to create a new experiment that could be used for the course in the future [5]. While these proposals are not actually constructed during the semester they are proposed, at least six of them have been built to use for the course over the years.

The ChE 426 course is the second unit operations laboratory course, and it has many of the same safety implementations in it as the first. The biggest difference is that in this course, at least one of the experiments conducted is considered a "Problem-Based Learning" experiment, or an open-ended experiment. In it, the students are given a general problem statement and they need to design an experiment using equipment available in the lab to test the hypotheses that they develop about that problem. In doing this, they must present their design and include potential hazards of the tests that they plan to conduct. In doing this, they are forced to apply inherently safer design techniques and consider the mitigation steps needed to keep themselves safe in the lab while conducting their experimental tests.

ChE 421: Process Dynamics and Control

The Process Dynamics and Control course falls into the senior year in the program of study for most students. In this course, the integration of process safety and the course subject matter becomes very apparent to the students. Once students have learned about control algorithms and logic, the equipment used for these systems is discussed. The action required for valves, such as "fail-open" or "fail-closed" is discussed, as is the selection of these items for specific process requirements. Other physical equipment is explained, including check valves, redundant measurement devices, and safety instrumented systems. On the software and control logic side of things, students are introduced to the concepts of reverse and direct acting controllers and how they are chosen for safety function, interlocks and their purpose, and PLC's that can be used for safety functions. Finally, students learn the "probability" piece of risk assessment as they learn the basics of fault tree analysis. This includes defining the logic diagrams used in a fault tree, a discussion of failure rates of different types of equipment in the control system, and how to minimize the overall probability of an event. They also learn about the application of probability solutions to various kinds of networks and discuss the assumption of independence and analysis when independence is not reasonable.

ChE 430: Chemical Process Design

A process design course is a logical place to incorporate process safety principles, and in fact, many chemical engineering programs do this as their primary method for addressing the ABET requirements for hazard awareness [2] [3]. At ISU, we include process safety considerations as a part of the process design course, which is taken during the final semester of study for most students. Throughout the course, the concepts of inherently safer design are discussed as part of the design process. Special consideration is also given to certain process conditions that can be hazardous, including high pressures and temperatures. There are several lectures included in the course to address specific concerns relating process design and process safety.

The first special lecture is about materials of construction and how to choose them for process fluid compatibility and for process conditions. This includes a discussion of corrosion chemistry and corrosion mechanisms and addresses the potential for catastrophic failures of systems that can result from corrosion.

Two lectures are dedicated to covering safety basics related to design of processes. In these lectures, an overview of risk assessment and hazard identification methods, such as bow-tie models and HAZOP studies, is given. The basic definition of risk is used, where risk is considered to be the probability times consequence of an accident. In order to evaluate probability further and apply it to their projects, students receive an introduction to failure and reliability of process components. They learn about the "probability of failure on demand" for safeguards and the use of multiple mitigation devices to reduce overall risk. A high-level overview of the OSHA PSM regulation is presented, and its effect on different parts of processes is discussed. Environmental regulations and their application to process design are also reviewed, including RCRA, TSCA, and PHMSA. There is an overview of the types of statistics used to measure performance related to safety (injury/fatality statistics, etc.).

During the semester, students work in teams to complete two specific design projects. Each of these requires some analysis and discussion of health, safety, and environmental considerations. In their designs, they must identify any conditions of special concern and appropriate safety systems that should be used to mitigate the risks. They also must specify appropriate materials of construction, evaluate NFPA ratings and process stream hazards, and propose process waste stream management or treatment that is necessary. In addition, once they have identified the process hazards, they must perform a HAZOP analysis and construct a bow-tie plot for the component or process stream that they have deemed is the greatest threat in the process.

ChE 420: A Dedicated Course for Chemical Process Safety

Along with the course modules that were being developed throughout the curriculum, a new dedicated process safety course was proposed in 2010 and was accepted to be offered for the first time in Fall 2012, as a three-credit-hour course. This was intended to be an elective course initially, for senior-level students who planned to work in industrial positions after graduation. However, after the first semester offering, the department decided to make it a graduation requirement based on feedback from students and to ensure that process safety and hazard awareness was taught to all students, in accordance with the ABET requirements that took effect in 2012 [1]. Every required course is offered every semester at ISU, so that students are encouraged to take co-op and internship opportunities without worrying about how their educational progression will be affected. As a result, because the process safety course became required for all students starting in the 2013-2014 catalog, the course has been offered every semester since Fall 2014. Enrollment data for this course are shown in Figure 1. To date, 1206 students have been through the course.

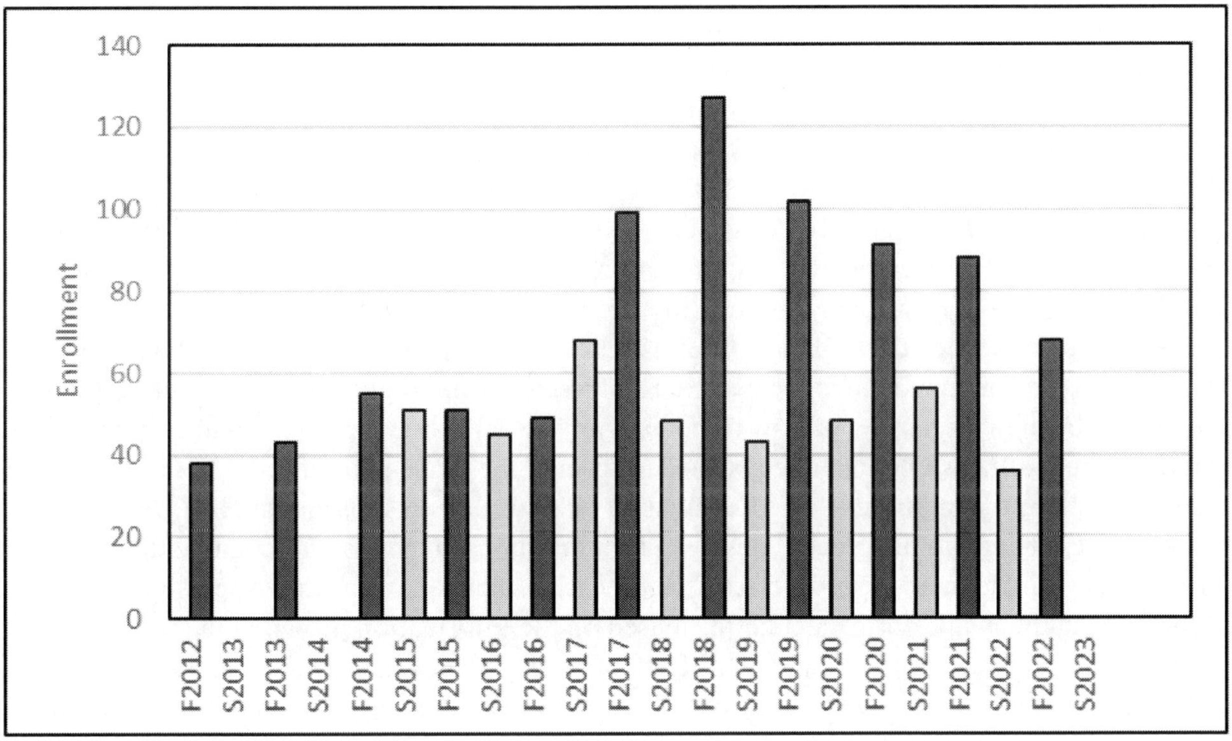

Figure 1: Enrollment in ChE 420, the dedicated process safety course at ISU, by semester since offerings began in Fall 2012. Beginning in the 2014-2015 school year, it has been offered every semester because it became a graduation requirement for students entering under the 2013-2014 catalog.

The course is titled "Chemical Process Safety" and follows the text written by Dan Crowl and Joseph Louvar, now in its fourth edition [6] . Most of the chapters in the book are covered over the course of one semester, but some are not covered in detail because of time constraints. The learning outcomes for the course are as follows:

1. Understand government regulations applicable to process safety
2. Understand toxicology
3. Develop and apply source models to estimate release rates and quantities
4. Develop and apply dispersion models to estimate concentrations of hazardous materials
5. Identify, formulate and solve engineering problems
6. Understand the fire triangle and how to prevent flammable mixtures
7. Identify and address reactive hazards
8. Understand relief system design
9. Understand and apply the risk assessment process
10. Understand the concepts of inherently safer design
11. Demonstrate the ability to work effectively as a member of a team
12. Demonstrate knowledge through written communication
13. Recognize the need for lifelong learning through the evaluation of a process in a hazard analysis
14. Use computational tools

Over the course of the semester, case studies are used regularly in order to illustrate principles from the different subjects as they are being studied. A number of these come from the CSB video library [7], but some other sources are also used. When choosing a case study to evaluate, a number of factors are considered, including time constraints, format of the case study materials, and intended activity in the classroom. For example, when introducing explosions, there are videos available on YouTube that have good explanation of how a BLEVE occurs [8], and that depict visually very well how a shock wave travels across a desert surface [9]. These are simple videos to illustrate a specific point before we begin to study the material in class. Some case studies involve longer videos and reading materials about specific cases, such as the T2 Laboratories explosion or the Williams Olefins reboiler overpressure [7]. In the T2 case, the video and CSB recommendations are used to reinforce the importance of understanding reactive hazards as we begin discussion of how to obtain and use calorimetry data to characterize reactions. The Williams Olefins video is used to discuss the placement of relief devices in a process as well as the importance of re-evaluating hazards during subsequent HAZOP evaluations, even though the process has been running for many years. Other case studies are more in depth, and homework or in-class assignment problems have been written so that students use the case study and perform calculations based on the information provided in the accident reports. And example of this is the Millard Refrigerated Services incident in Mobile, Alabama [7]. Using this accident investigation information, students are given a problem to work in teams to estimate the source model release area, and to perform dispersion modeling to estimate concentrations of ammonia released in the area around the facility.

In addition to regular assessments including homework, exams, and quizzes, the final assessment in the process safety course is a comprehensive project that is centered around performing a HAZOP/LOPA analysis of risk for a specific process. Processes used for this change each semester and have included several of the "Preliminary Design of Fifteen Chemical Processes" in Turton's *Analysis, Synthesis and Design of Chemical Processes* [10], process designs provided by alumni, and a pilot-scale process used at a university research facility. In the project, teams of students apply the HAZOP and LOPA methods for hazard identification and probability analysis to determine overall risk using a risk matrix. The project has been done as a mock Process Hazard Analysis and as a hazard analysis for Management of Change. In addition to the hazard analysis, students are given specific questions to answer and calculations to perform so that the instructor can perform a comprehensive assessment of their learning from the entire semester.

Future Plans

It has become apparent that the field of process safety is continuously developing and improving, and there is much more that could be taught to undergraduate students to give them more depth of understanding. The modules used throughout the curriculum as well as the dedicated safety course lay the groundwork for basic understanding and help the students prepare for jobs in process industries as well as research, but we wish to offer them the opportunity to develop deeper understanding in specific areas of process safety. To that end, a new elective course is in development now. It will focus on fire and explosion dynamics first, and then a number of loss prevention concepts are under consideration. The first offering of this course will be Spring

2023, and it will be offered for a few semesters as a ChE 490 – Independent Study – course. The purpose of this offering is to refine the syllabus. The instructor will teach the fire and explosion dynamics as a standard class offering for the first part of the semester, and then groups of students will be allowed to choose an area of interest to perform research and develop course content that could be used in a future three-credit-hour course. This will be actively promoted to both chemical engineering and mechanical engineering students, as it will benefit any engineer who plans to work in a process facility. The dedicated process safety course will not be a prerequisite for the new elective course, but the two will complement each other and students who take both with be better prepared for a career as a process engineer or safety professional as a result.

Summary/Conclusions

The process safety curriculum at Iowa State University has been described. It has evolved over the past decade to include modules for several different core chemical engineering courses as well as a dedicated process safety course. Future plans to add a process safety elective course to give students more depth of understanding were shared. The approach taken at ISU is one that could be followed by other universities and is one way to ensure that chemical engineering curriculum covers the hazard awareness requirements set forth by ABET and AIChE.

References

[1] ABET, "Criteria for Accrediting Engineering Programs, 2022-2023," 2021. [Online]. Available: https://www.abet.org/accreditation/accreditation-criteria/criteria-for-accrediting-engineering-programs-2022-2023/. [Accessed 7 November 22].

[2] S. J. Dee, B. L. Cox and R. A. Ogle, "Process safety in the classroom: The current state of chemical engineering programs at US Universities," *Process Safety Progress,* vol. 34, no. 4, pp. 316-319, 1 December 2015.

[3] A. J. Pintar, "Integrating process safety into the unit operations laboratory," in *ASEE Annual Conference Proceedings*, Seattle, WA, 1998.

[4] DuPont, "At our core: the historical origins of safety at DuPont," 22 May 2019. [Online]. Available: https://www.dupont.com/news/safety-at-our-core.html. [Accessed 9 November 2022].

[5] S. D. Loveland and K. L. Stoner, "Design Projects in the Unit Operations Laboratory," in *AIChE Annual Meeting Conference Proceedings*, Phoenix, AZ, 2022.

[6] D. A. Crowl and J. F. Louvar, Chemical Process Safety, 4th ed., Boston: Pearson, 2019.

[7] U.S. Chemical Safety and Hazard Investigation Board, "CSB Video Room," [Online]. Available: https://www.csb.gov/videos/.

[8] VideoSpikes, *BLEVE (Boiling Liquid Expanding Vapor Explosion) - Demonstration How it Happens Training Video,* 2010.

[9] "Shockwave, Season 1 Episode 2 - "The Pepcon Disaster"," Cheri Sundae Productions, 7 December 2007. [Online]. Available: https://www.imdb.com/title/tt1195340/?ref_=ttpl_pl_tt.

[10] R. Turton, R. C. Bailie, W. B. Whiting, J. A. Shaeiwitz and D. Bhattacharyya, Analysis, Synthesis, and Design of Chemical Processes, 4th ed., Prentice Hall Int, 2014.

239e How We Designed Our Safety Program to Improve Safety Education and Process, and to Satisfy Aiche Program Criteria

George Prpich, Ron Unnerstall

Process safety is a critical element of a Chemical Engineer's education and has become a valued engineering skill in the hiring process. Beginning two years ago, we began a self-examination of our safety program and processes. We looked specifically at how safety procedures were being applied in research (e.g. graduate and undergraduate) and how safety concepts were being taught within the undergraduate program curriculum. We learned that our safety program was in need of a refresh, for example, our departmental safety culture had become heterogenous and our undergraduate safety program lacked context and cohesion. Over the past two years, we have addressed these issues by implementing a number of changes to our safety program both inside and outside of the classroom. Within the classroom we have designed, implemented, and assessed several new educational interventions. For example, we created a new dedicated process safety design module (1-credit) that is coupled with the 4th year Capstone experience. We also integrated novel hazard assessment and process safety pedagogy into our Unit Operations Labs (e.g. safety moments, incident reporting). In addition to these core elements, we also created a new undergraduate process safety elective course (3-credits), and initiated an undergraduate process safety research program. Outside of the classroom, opportunities for improvement in safety processes were initially informed by a safety culture survey that we used to benchmark safety practices and behaviors across the department. Results from this study led to a refresh of our safety procedures and protocols, e.g., revised safety manual and the development of an incident reporting mechanism. Integrating change both inside and outside the classroom has led to an improvement in the quality of safety being practiced and discussed in research labs and classrooms alike.

In this presentation, we will share our journey from two years of effort spent updating our safety processes. We will present data and discuss insights gathered from student feedback about perceptions of safety, safety incident reporting in the department, and from safety moments presented by both graduate and undergraduate students. We will also discuss our strategy for continuous improvement (e.g. strengthen safety culture) and share why we believe our integration of safety processes inside and outside of the classroom is a great approach for satisfying the AIChE Program Criteria.

239f Integrating Process Safety with a Unique Industrial Perspective into the Curriculum at Kansas State University

Jennifer L. Anthony, Md Uddin, Andrew Duncan, Christopher Frampton, Chris Aiken

At Kansas State University (K-State) we address the design, analysis and control of processes, and associated hazards curriculum ABET requirement extensively through a multi-pronged approach during the students' senior year. Our approach includes one process safety course, one controls course, two design courses and a partnership with Cargill, Inc that provides every design team with an industrial process safety coach in addition to their design project mentor. Students complete three separate design projects over the course of the year which are evaluated using the same rubric, allowing for clear assessment of our outcomes. We will discuss the impact on student performance and modifications we have made since these changes were first implemented in 2018.

239g Successful Process Safety Instruction in Two Capstone Design Courses at a Hispanic-Serving Institution

Matthew Alexander, Joseph Amaya

The senior design courses of our chemical engineering curriculum serve as a transition for students completing their undergraduate studies and preparing to enter the engineering workforce. Therefore, the capstone courses are a logical place for process safety instruction, since this topic fits well with the course emphasis on process design and the senior capstone project experience. The instructor draws on his 25 years of industrial experience for teaching and mentoring students on various process safety topics, in order to satisfy the ABET curricular requirement for process hazards and process safety instruction. The topics presented in several lectures in both the first and second semesters of the capstone design course sequence include chemical hazards, process safety analysis, industry process safety management systems mandated by OHSA regulations, and established safety tools such as the HAZOPs analysis. This talk will present the instructors' comprehensive approach to process safety instruction, and the student level of learning achieved in this topic area.

Many of our graduating seniors find employment in one of the refineries and chemical plants that are prevalent in the region near our university. Nearly half of our students come from the local area where refinery and process chemical plants are numerous, while a majority of the remaining students come from a region of our state where there are virtually no refinery or process chemical facilities. One of the challenges we face in this task is adapting the instruction to this diverse student body, half of whom have been exposed to process safety issues from internships or talking with friends or relatives in the industry, and half whose exposure to safety has been nothing more than the high school or college chemistry lab class. The various facets of the process safety instruction which we have developed and incorporated into our senior design course sequence is intended to provide our students with a well-rounded overview of safety topics to prepare them for employment in the process operating environment. First off, a distinction is made between personal safety and process safety, realizing that chemical engineers will be dealing with both of these when working in a process environment. Secondly, chemical and refinery accidents investigated by the Chemical Safety Board are presented and discussed in order to impress upon the students the very serious nature of process safety and process hazards in their future work environment. Third, an overview of OSHA process safety management (PSMs) systems that refineries and chemical manufacturers are required to develop and implement is explained. Finally, tools used in PSMs, such as HAZOPs analysis and PHAs, are discussed and utilized in instruction. The students respond well and usually score high on related assignments, such as written technical memos and quizzes, on these various safety elements. From this, we judge that the process safety instruction is effective. Additionally, a complementary indication of success to date of our multi-faceted approach is the recognition of this safety instruction as a strength of our program by the ABET reviewer in a recent ABET site visit.

239h Integrating Process Safety across the Curriculum

Tracy Carter

Failure to prepare students with essential process safety knowledge and skills has led to catastrophic and financially devastating events.[1] To support college and university continuous curriculum improvement in process safety, ABET, requires evidence that students understand "the hazards associated with these processes" as part of chemical engineering program accreditation. [2] However, incorporating chemical process safety outcomes across the curriculum is challenging because creating change in higher education is difficult. Barriers to changing curriculum include "a lack of time, instructional challenges, and a lack of autonomy". [3] Leveraging drivers for change can help. These drivers include "expanding on current practices, encouraging collaboration, and common objectives for improving teaching and assessment".[3] One change model has been suggested to create change in higher education. This model includes four strategies: "Developing a Shared Vision, Enacting Policy, Disseminating Curriculum and Pedagogy, and Developing Reflective Teachers".[4]

To create change, process safety outcomes needed to be defined. In 2010 the Safety and Chemical Engineering Education (SAChE) Committee in collaboration with faculty and industry leaders, developed eight outcomes. These eight outcomes include an understanding of past events, the hazards and impacts of chemicals and releases, as well as understanding how to assess the hazards and risks associated with a process and to design or mitigate against them.[5] Using the proposed change model and leveraging drivers enabled the integration of chemical process safety outcomes into eight core courses across the curriculum.

1. US CSB. (2022). T2 Laboratories, Inc. Runaway Reaction. https://www.csb.gov/t2-laboratories-inc-reactive-chemical-explosion/.
2. ABET. (2022). Accreditation Policy and Procedure Manual. https://www.abet.org/accreditation/accreditation-criteria/accreditation-policy-and-procedure-manual-appm-2020-2021/.
3. Shadle,S.E., Marker,A., Earl,B., International Journal of STEM Education 2017,4,(1),8.
4. Henderson,C., Beach,A., Finkelstein,N., Journal of Research in Science Teaching 2011,48,(8), 952-984.
5. SaChE. (2022). The Safety and Chemical Engineering Education Program. https://www.aiche.org/ccps/community/technological-communities/safety-and-chemical-engineering-education-sache/sache-products.

188a Evidence-Based Inclusive Teaching Strategies

Stephanie Velegol

There is an increased interest in using teaching strategies that will attract and retain a diverse set of students in chemical engineering. The purpose of this talk is to share an overview of evidenced-based inclusive teaching strategies. These strategies are broken down into three main categories that have been shown to increase student retention and motivation for learning: 1. Capacity and self-efficacy, 2. Interest, 3. Belongingness and support. For example, Canning *et al.* showed that faculty with a growth mindset lowered the achievement gap between white/Asian and underrepresented students (Canning 2019). Including formative assessments instead of summative assessments (e.g. exams that "sort") had a large impact on students with low social economic backgrounds and no impact on those with high social economic backgrounds (Smeding 2013). In another publication, Stephens *et al.* showed that a one-hour intervention using difference education resulted in significant GPA gains for first generation students but not for continuing education students (Stephens 2014). An example of difference education is a first-generation student sharing that it was important that they find an advisor since their parents didn't have college experience. These and other publications serve as a platform on which you can build an inclusive experience for a wide variety of students in your program.

Canning, Elizabeth A., Katherine Muenks, Dorainne J. Green, and Mary C. Murphy. 'STEM faculty who believe ability is fixed have larger racial achievement gaps and inspire less student motivation in their classes.' Science advances 5, no. 2 (2019)

Smeding, Annique, Céline Darnon, Carine Souchal, Marie-Christine Toczek-Capelle, and Fabrizio Butera. 'Reducing the socio-economic status achievement gap at university by promoting mastery-oriented assessment.' PLoS One 8, no. 8 (2013): e71678.

Stephens, Nicole M., MarYam G. Hamedani, and Mesmin Destin. 'Closing the social-class achievement gap: A difference-education intervention improves first-generation students' academic performance and all students' college transition.' Psychological science 25, no. 4 (2014): 943-953

188b Increasing Accessibility for Visually Impaired STEM Students through a New Design and Rapid Prototyping Course

Vance Jaeger, Adam Stockhausen

Blind and visually impaired (VI) students often struggle to understand STEM concepts due to a lack of available teaching materials. Commonly, K-16 STEM lessons are presented using images, videos, graphs, and visual demonstrations. However, these traditional visual methods are unable to reach VI students, and STEM instructors who do not have experience teaching VI students struggle to convey ideas in a relatable way. To address the lack of educational aids for VI students, we have developed a new design and prototyping elective for upper-level undergraduate and graduate students within Chemical Engineering departments. In the new course, university students use computer-aided design (CAD) and 3D printing to produce educational models specifically for grade 9-12 VI students in STEM classrooms. Models are designed specifically to provide tactile cues to understand difficult STEM concepts and to be used completely without the need for vision. With feedback from the high school science instructor at the Kentucky School for the Blind (KSB), students at the University of Louisville worked throughout Spring 2022 to iteratively design and improve models for grade 9-12 classrooms. The university course curriculum covers not only the skills needed to design and produce models (e.g., CAD, 3D printing, polymer science, polymer processing) but also gives the students a deeper understanding of the need to consider accessibility and inclusion in their designs and in their careers.

In the inaugural year of the course, thirteen university students were enrolled, ranging from juniors to Ph.D. students and representing three engineering departments. The university students worked both individually and in teams to produce educational models covering STEM concepts from the state's K-12 academic standards. In total, seventeen designs were produced to aid in instruction of blind and VI students at KSB. Major concepts covered with these models include but are not limited to trends in electronegativity and atomic radii on the periodic table, dominant and recessive heritability, plant life cycles, DNA structure, layers of the earth, electromagnetism, and the central limit theorem. Students have chosen to upload their CAD models for free to public repositories to allow for STEM instructors across the world to benefit from their efforts. We are in the process of compiling tutorials and guides for K-12 instructors to 3D print these models for their own classrooms. To replicate this course at other universities, the cost of equipment and materials is about $2000 for a class of 10-20 university students, which makes the project accessible to all engineering departments. The course curriculum, tutorials, and suggestions for integrating the models into the K-12 curriculum will be provided online for free. Importantly, the course concept can be easily replicated, modified, and improved to address the needs of students outside grades 9-12, students outside STEM disciplines, and students with other special learning needs.

188c 10 Tips to Make Your Course More Accessible and Inclusive to Disabled Students

Mariah Arral

Accessible chemical engineering education is hindered for disabled students by systemic ableism. Section 504 of the Rehabilitation Act of 1973 mandated equal access to postsecondary institutions for disabled people. Sadly, since the passing of the bill, there has been pushback. Laws, research, and activism have allowed improvement, but there are still many steps we must take to further the accessibility of education and inclusion of disabled students. Research has shown that disabled students face systemic barriers ranging from general lack of support to negative views from faculty members. Educators identified limited resources as a barrier to supporting disabled students. There is a clear need to ensure faculty have the resources to adapt their classrooms to an accessible format. Here I will provide 10 guidelines to help ensure a classroom is more accessible and inclusive to disabled students. The guidelines are as follows: 1) Instructor and course specific accessibility and accommodations statements, 2) Meet with students privately about their accommodations, 3) Do not ask students to disclose their disabilities, 4) Consider the implications of language used, 5) Apply principles from Universal Design for Learning in the classroom, 6) Proactively provide accessible materials, 7) Upload notes and slides for class and/or recordings, 8) Remove attendance requirements, 9) Check-in with students on how accommodations are working in your class, 10) Include content from those in the disability community in the curriculum. This presentation is supported by disability and education literature. Adopting these guidelines is just the first step in starting to create a more inclusive and accessible educational environment. In conclusion, acquiring accessible education for all is within our power, but resources for 1) disabled individuals and 2) educators are needed.

M.L.A. was supported by an NSF Graduate Research Fellowship Program award (number DGE1745016)

188d Understanding Barriers to Mental Health Related Help-Seeking in Undergraduate Engineers

Courtney Wright, Lucy Hargis, Melanie Miller, Ellen Usher, Joseph Hammer, Sarah Wilson

An increasing concern on college campuses is the prevalence of mental health disorders. Within engineering, there has been an increase in the amount of research conducted on characterizing student mental health and understanding the influence of engineering culture on mental wellness. Through analysis of national data from the Healthy Minds Study, it has been shown that distressed engineering students are less likely to have received professional help compared to their peers. This signifies a treatment gap for engineering students, meaning that only a portion of those in mental health distress seek professional treatment. Addressing the treatment gap can improve the prognosis of mental health problems and reduce progression to more chronic or severe disorders. Through this talk, I will present the results of a qualitative study aimed at better understanding the beliefs that engineering students have about seeking help for mental health concerns. In particular, I will discuss perceived barriers and facilitators to help-seeking and the three themes identified in the study, including challenges in navigating the mental healthcare system, the perception that help-seeking is a sacrifice of time and academics, and the importance of mental health culture in engineering. Results of this work will help provide guidance for intervention development to improve mental health related help-seeking among undergraduate engineering students. This is important for creating an equitable engineering learning environment that is supportive and inclusive of those in mental health distress.

188e A Diversity Index to Assess College Engineering Team Performance and Tracking Insitutional Changes on Diversity

Joaquin Rodriguez Alonso, John Keith, April A Dukes

We have developed a Diversity Index (DI) to better quantify the impact of eight traditionally underrepresented demographic categories in chemical engineering ("Women," "Non-Binary," "Black or African American," "Hispanic or Latino/a," "Asian, American Indian or Alaska Native, Middle Eastern, or Native Hawaiian or other Pacific Islander (Asian/other for a short version)," "LGBQT+," "Disabilities," and "First Generation"). The complexities of conceptualizing and measuring diversity have been thoroughly examined in the literature. The dominant methods to estimate diversity have been reviewed. The proposed Diversity Index offers advantages in terms of the simplicity of calculations, while providing a comprehensive structure to analyze the contributions of underrepresented groups. This Diversity Index can then be used to track institutional changes in the inclusion of diversity at any level (college, school, and department). Moreover, the index is particularly valuable to address the configuration of small teams (i.e., 4-6 members) in engineering courses, providing a quantitative tool for faculty to assess the impact of diversity on teamwork performance.

The Diversity Index has been developed and tested throughout the last four years (191-2022) with nine chemical engineering Pillar courses (Process Reactive Engineering, Process Control, and Process Design). The evaluation includes 410 students, grouped in 76 self-selected teams, performing over 45 technical projects, and 108 outreach projects. Results confirm the bias towards more uniformity when teams are left to the self-selection of members. However, this index provides quantitative assessment on the spread of diversity in those groups, that faculty can use to establish constraints or reward diversity in teams. Moreover, results show a promising method to correlate diversity and team performance. In addition, this index is open to add other categories that could prove increasing potential for correlation with team performance, and includes a tuning parameter to be adjusted by research to adapt to various societal contexts. In this regard, this index opens many opportunities for specific research on tracking institutional changes on diversity, admission and retention of underrepresented groups, and team performance in assignments and projects.

Diversity is currently sought as a necessary component of engineering education, required for the future workforce, faced with dramatic demographic changes across the world. However, diversity continues to be a much debated and often confusing term, where the relation between team diversity and performance has not shown conclusive results, and very often, contradictory trends. The proposed Diversity Index can be developed into a valuable tool to assess quantitatively and qualitatively all these significant issues in engineering education.

188f Best Practices for the Advancement of Diversity, Equity, and Inclusion in Chemical Engineering Education

Reginald Rogers Jr., Todd Pagano

There is a problem with the low number of students from underrepresented groups entering the field of chemistry and chemical engineering, persisting to degree completion, and entering the workforce. These are ethical and practical concerns when we lose out on diversifying the chemical sciences/engineering and miss benefitting from the greater contributions of diverse thinkers working together in the classroom and workplace. The principles of diversity, equity, and inclusion (DEI) are critical in resolving current inequities. Continuous reflection is a critical component of the improvement process and catalysts along the path to DEI include understanding value and embracing change. Strategies for making students feel a sense of belonging in the course, research laboratory, and discipline include ensuring the educational climate is safe, accessible, and respectful; making the learning experience relatable and providing access to role-models; and providing research experiences and exposure to the field. We discuss strategies that individuals can utilize to help lower the activation barrier needed to realize DEI on the large scale.

188g Enhancing Chemical Engineering Identity in Young Women with a Biomedical Polymer Outreach Activity

Jessica Torres, Julie C. Liu

Women obtain a disproportionately low percentage of chemical engineering degrees compared to the general population. A lack of understanding of what engineers do and misguided emphasis on math and science instead of problem solving are cited as reasons why women choose not to pursue engineering degrees [1]. Programs that introduce students to engineering are useful tools for altering preconceptions about engineering. They can also serve as a useful tool for students to try out a field in a low-risk setting. We developed a chemical engineering outreach activity for young women focused on biomedical applications of polymers to introduce non-traditional chemical engineering fields. In this work we evaluated the impact of the activity on young women's interest in chemical engineering and on their likelihood of pursuing a chemical engineering degree.

The outreach activity was a session for Purdue Women in Engineering Program's Introduce a Girl to Engineering Day, a free day-long event for high schoolers in 9th and 10th grades. The session was composed of a short presentation introducing chemical engineering and polymers. In the activity, students explored the potential for the super absorbing polymer, sodium polyacrylate to function as a hydrating wound patch for burn victims. The activity was broken into three challenges. The first challenge showed students two extreme polymer to water ratios; the second challenge encouraged students to optimize their own ratio for hydration and stiffness; and the final challenge was to translate the material for industrial use. A survey was given to students before and after the chemical engineering activity. Students were asked to access their agreement with statements about their opinions of their technical abilities (academic identity), career interests (occupational aspirations), and thoughts of the chemical engineering field (occupational identity). Figure 1 lists the questions asked in both the pre-survey and the post-survey. The two biggest increases in agreement from the survey were in the statements "I am interested in chemical engineering," and "Chemical engineers help people." Agreement with all statements increased except the statement "I want to help people in my future profession," which remained unchanged.

The differences in occupational aspirations and occupational identity were also measured as a result of the activity (Figure 2). This comparison was intended to determine if there was a mismatch between what students want to do in their future professions and if they think chemical engineers do those things. When asked before the activity if they wanted to help people, work with others, and be creative, most students strongly agreed, but fewer students thought they could do these things as chemical engineers. After the activity, the mismatch decreased or flipped.

Overall, this biomedical polymer activity can increase young women's understanding of chemical engineering and align their career values with the chemical engineering field. This activity introduces chemical engineering from a broader lens and emphasizes creativity, teamwork, and helping others as components to the chemical engineering discipline.

References: [1] Isaacs B., *J Prof Iss Eng Ed Pr*, 127, 2001.

IGED Chemical Engineering Session Evaluation

On a scale of 1 to 5, please indicate the extent to which you agree with the following statements (please circle one):

	Strongly Disagree	*Disagree*	*Neutral*	*Agree*	*Strongly Agree*
I am interested in chemical engineering.	1	2	3	4	5
I think I can be a chemical engineer.	1	2	3	4	5
I am good at solving problems in math and science.	1	2	3	4	5
I want to help people in my future profession.	1	2	3	4	5
I want to work with others in my future profession.	1	2	3	4	5
I want to be creative in my future profession.	1	2	3	4	5
Chemical engineering is a profession where you get to help people.	1	2	3	4	5
Chemical engineering is a profession where you get to work with others.	1	2	3	4	5
Chemical engineering is a profession where you get to be creative.	1	2	3	4	5

Figure 1. Survey statements asked before and after the chemical engineering activity. Responses were scored from 1 (strongly disagree) to 5 (strongly agree) from each student who participated in the activity.

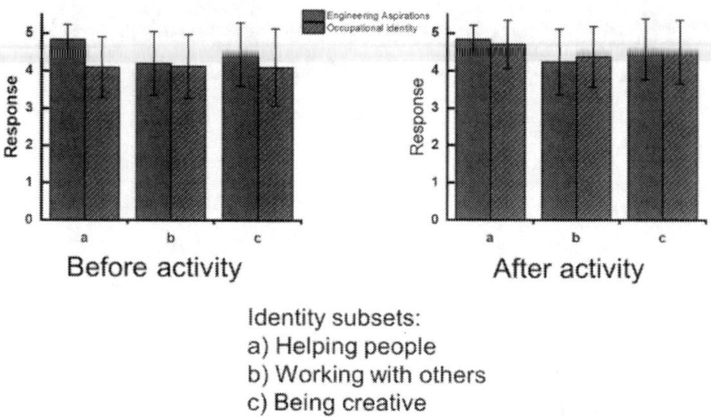

Figure 2. Responses comparing occupational aspirations to occupational identity before and after the activity. Identity aligned more with aspirations after the activity. Questions of similar structure (helping people, working with others, being creative) were compared. Responses were scored from 1 (strongly disagree) to 5 (strongly agree) from each student who participated in the activity.

188h Why Incorporating Community and Global Issues into the Classroom Matters

Joni Lakin, Edward W. Davis, Virginia Davis

In the fourteen years since the National Academy of Engineers started "changing the conversation" through its engineering messages campaign, there has been increasing emphasis on highlighting the vital role engineers play in solving grand challenges such as access to clean water and more affordable renewable energy. This emphasis on how engineers solve problems, instead of creating them, echoes calls for action from other agencies such as the United Nations, the American Chemical Society, and the Royal Society of Chemistry. However, pre-college students still tend to believe that engineering is an individualistic, status-driven profession and not a collaborative, altruistic one. As a result, students from demographics that tend to favor "helping professions," may find pursuing an engineering degree incompatible with their core values. For those students that do enroll in engineering programs, this belief can either be reinforced or ameliorated by how engineering is framed in their classes and extracurricular activities such as the Grand Challenge Scholars program and Engineers Without Borders. This talk summarizes eight years of research by the authors on how framing engineering as an altruistic profession affects engineering identity development. It places a particular emphasis on using social congruity theory to help understand the engineering attitudes of female and underrepresented minority students. In addition, it provides examples of activities that be used to teach both core chemical engineering concepts while highlighting engineers' vital role in making the world a better place. In total, the results of this research highlight that intentionally framing engineering as altruism is compatible with teaching core concepts and reinforces learning by helping students align pursuing an engineering career with their core values.

186a Fostering Inclusivity By Linking the Implementation of Computational Simulations to Student Learning Styles

Ifeoluwa Babalola, Victor Ugaz

Computational simulations enhance undergraduate engineering education by enabling complex phenomena to be presented in a visual way that appeals to a variety of learning styles and strengthens the link between theory and applications. Embedding visualization throughout the curriculum is also vital to promoting inclusive learning because students whose learning styles place them at the greatest risk for retention receive the most significant positive impact. But simulations are often ineffective because they employ complex computational tools and are implemented without considering the link to student learning style. Here we describe a study aimed at understanding this link by assigning students in an undergraduate chemical engineering transport phenomena class to perform a computational simulation exercise in teams grouped according to learning style. Student learning styles were classified according to the Myers Briggs Personality Type Indicator assessment. The simulation exercise was presented using two computational tools of varying complexity, STAR CCM+ and openFOAM. We describe the initial results of a pilot study and identify insights likely to maximize the impact of simulation-based instruction, particularly for students with at-risk learning styles.

186b Incorporating Circular Economy in Chemical Engineering Education: A Multidisciplinary Approach

Mamoun Al-Rawashdeh, Konstantinos E. Kakosimos, Peter Martin, Joselia Neves, Christine Schiwietz, . Savanid (Nui) Vatanasakdakul, Evren Tok, Dhabia Al-Mohannadi

Students entering universities, will have to live in a world beyond the Sustainable Development Goals (SDGs). Chemical engineers of the future will have to design process and products that fit within sustainable production and consumption (SDG: Goal 12). It is no longer an aspiration, but a reality that achieving economic growth and sustainable development requires that we urgently reduce our ecological footprint by changing the way we produce and consume goods and resources. This requires engineers to understand social, economic, and environmental needs of product including the end of life. Engineering education focuses on technical feasibility of a product, and often lacks the accounting of human interactions with consumer targeted products. PRODUCT X: Innovations for Future Economies Bootcamp is a ground-level initiative of faculty from multiple Universities located in Qatar's Education City Universities (https://www.qf.org.qa/education/education-city) to engage students with the scope, methods, and principles of interdisciplinary innovation responses to complex issues. It brings together engineering students with students from different majors such as graphic design, business, political science, culture, and sociology across five universities. This bootcamp provides a multiversity platform for students to collaborate in innovating product concepts and system scenarios designed to facilitate socio-technical transitions to future economies that are more sustainable and equitable. This "incubation style" puts together multidisciplinary teams with faculty mentors to form creative, entrepreneurial, and sustainable ideas. The students focused on eliminating or replacing single-use plastic containers products with a sustainable solution. Finding a sustainable solution is complex task that involves addressing consumer behavior, regulations, costs, and footprints. The aim is to promote 'circular' thinking in development and consumption, enable students to gain communications skills with other disciplines and provide a holistic understating of sustainable design.

186c Modernizing Satisfaction of the ABET/Aiche Process Safety Criterion through an across-the-Curriculum Approach

Matthew Cooper

In collaboration with AIChE, ABET specifies that that the curriculum of chemical, biochemical, biomolecular and similarly named engineering programs must include a unique program-specific criterion related to process safety. However, safety training of undergraduates is far from standardized, with one 2015 survey indicating that 23% of programs have a dedicated process safety class, implying that the balance of programs integrate this education into other core classes in their curricula. The latter strategy will be discussed in this presentation. NC State's ChE program leverages SACHE modules, the Contents Under Pressure process safety decision making video game, and traditional safety-related problems to address the criteria. Discussions of assessment strategies and how the approach addresses the safety-related program criterion will be presented.

186d Implementation of Molecular Biology and Cell Line Development Courses for Future Engineers in Biopharmaceutical Industry.

Driss Elhanafi

The success of the biotechnology industry relies largely on a highly educated and skilled workforce. Chemical engineers in careers related to the manufacturing of biopharmaceuticals engage with critical aspects in diverse upstream and/ or downstream processes such as the design and development of protein expression systems, or the production, purification and formulation of a therapeutic protein of interest. In order to empower graduate and undergraduate students with a strong understanding of the biomanufacturing workflow, we have designed two closely linked hands-on courses at North Carolina State University's Biomanufacturing Training and Education Center (BTEC) that emphasize the fundamentals of molecular biology and cell line development. The first course entitled "Molecular biology for biomanufacturing" exposes students to detailed principals and techniques allowing the comprehension of key steps in the establishment of a prokaryotic expression system. The lecture component covers fundamentals of genetics and physiology of prokaryotes, while the laboratory section of the course provides the students with a valuable opportunity to genetically engineer a bacterium for the production of a protein of interest. Experimental tasks include the cloning of the gene of interest on a specific expression vector, selection and screening of bacterium with the recombinant vector, small scale fermentation and production, followed by identification, quantification and testing the activity of the produced protein using specific analytical methods. The second course "Cell line development for biomanufacturing" highlights key aspects in the generation of mammalian and yeast expression systems using two cell types that are widely used in the biopharmaceutical industry. We underline specific cloning and expression considerations that are pertinent to eukaryotic systems with special emphasis on cell culture techniques, metabolic engineering using CRISPR technology, cell banking and good manufacturing practices. After producing the same protein in all three expression systems, students appreciate the advantages and disadvantages of each cell type at different levels including the process time, productivity, cost and activity. Keeping in mind the applicability of the courses learning objectives to biomanufacturing, the lectures and laboratory material have been developed to nurture the concept of maximum productivity of a protein with the desired quality attributes, facilitating the downstream processing, and how to troubleshoot possible bottlenecks. In addition to students in chemical engineering these two courses serve a number of other graduate and undergraduate programs at NC state university, including those majoring in biomedical engineering, chemistry, and biochemistry. Further, we have established an intensive short version of the two courses, conducted over three days, to train scientists from biopharmaceutical industry. Details of the courses and their impact on the biomanufacturing industry will be discussed.

186f Chemeng Remote Experience Augmented through Technology (CREATE) Labs – Training Leaders of the Future for Industry 4.0

Vijesh Bhute, Umang V. Shah, James I. Campbell, Andrew Macey, Sampad Sengupta, Pavan Inguva, Jerry Y. Y. Heng, Clemens Brechtelsbauer

Practical hands-on education is an integral component of STEM degrees in higher education institutions. While the complexity and independence in experiment design and practical problem selection for teaching increases with each year, all experiments go through several common stages including conceptualisation, design, hands-on operation, observation, data analysis, and presentation. Several of these steps need to be repeated based on observations. While these traditional in-person labs offer valuable training for undergraduate students, it may not prepare them well for their future careers as practitioners in industry. Owing to technological innovations, industry has seen significant use of state-of-the-art technologies including smart devices and automation of several tasks which minimises the hands-on operation of equipment. This switch is one of the key features of Industry 4.0. Chemical engineers, therefore, are expected to be well versed not only in the subject knowledge, but also have skills to communicate effectively with technologies and in remote teams.

In this work, we offer insights and evidence for effectively transforming traditional labs for remote or hybrid delivery and preparing students for Industry 4.0. We first discuss the evidence-based approaches for transforming hands-on labs into virtual, remote, or hybrid operation to achieve desired learning outcomes without compromising on soft skills and student self-efficacy [1]. We categorise different approaches based on claimed learning outcomes, availability of resources, technology, scheduling, and cost factors.

Using a case study for a 3rd year experiment [2], we discuss the implementation of Chemeng Remote Experience Augmented through TEchnology (CREATE) Labs which is a novel approach to deliver labs remotely with optimum in-person presence. They ensure that the experiments can be conducted by hybrid operation and students have to communicate effectively in teams working in person and remotely. Implementation is augmented with the use of video-led instructions, 'digital-twin' simulations, Pan-tilt-zoom (PTZ) cameras, and Microsoft HoloLens devices. Microsoft Teams is used as a platform technology for data sharing and communication. All traditional experiments for the three undergraduate year groups have been successfully delivered using the CREATE Labs approach over the last two years without compromising on student ability to understand, analyse or apply concepts. CRATE Labs were able to deliver experiential learning with no student presence during academic year (AY) 2020-21 and with 50% student presence during AY 2021-22. Indeed, their skills have been considerably enhanced for their future work environment where remote communication and operation is the norm.

We gathered evidence from various sources including surveys, interviews, and performance metrics to assess the effectiveness of CREATE labs implementation during the AY 2020-21 [3]. The overall delivery of CREATE labs during AY 2020–21 was positively received by both students and GTAs. The majority of students (>70%) reported experiencing effective communication with team members and GTAs and there was a strong positive correlation between communication and confidence in applying engineering concepts in the labs ($\chi^2 = 79.96$; $p = 1.69 \times 10^{-10}$). 5–10% of students from all year groups reported that they disliked the lack of in-person activities. The majority (>90%) of GTAs assisting with experiments stated that they associated their role in the CREATE labs with that of a facilitator. There was a year-dependent response on perceived confidence in achieving learning objectives which is likely a result of multiple factors.

Finally, we reflect on our learning from implementations over the past two years and discuss a plan for preparing students for effective remote communication and gain experience in using smart technologies which are key components of Industry 4.0.

References:

[1] V.J. Bhute, P. Inguva, U. Shah, C. Brechtelsbauer, Transforming traditional teaching laboratories for effective remote delivery—A review, Educ. Chem. Eng. 35 (2021) 96–104. https://doi.org/10.1016/j.ece.2021.01.008.

[2] U. Shah, P. Inguva, B. Tan, H. Yuwono, V.J. Bhute, J. Campbell, A. Macey, C. Brechtelsbauer, CREATE labs – Student centric hybrid teaching laboratories, Educ. Chem. Eng. 37 (2021) 22–28. https://doi.org/10.1016/j.ece.2021.07.004.

[3] V.J. Bhute, S. Sengupta, J. Campbell, U. V. Shah, J.Y.Y. Heng, C. Brechtelsbauer, Effectiveness of a large-scale implementation of hybrid labs for experiential learning at Imperial College London, Educ. Chem. Eng. 39 (2022) 58–66. https://doi.org/10.1016/j.ece.2022.03.001.

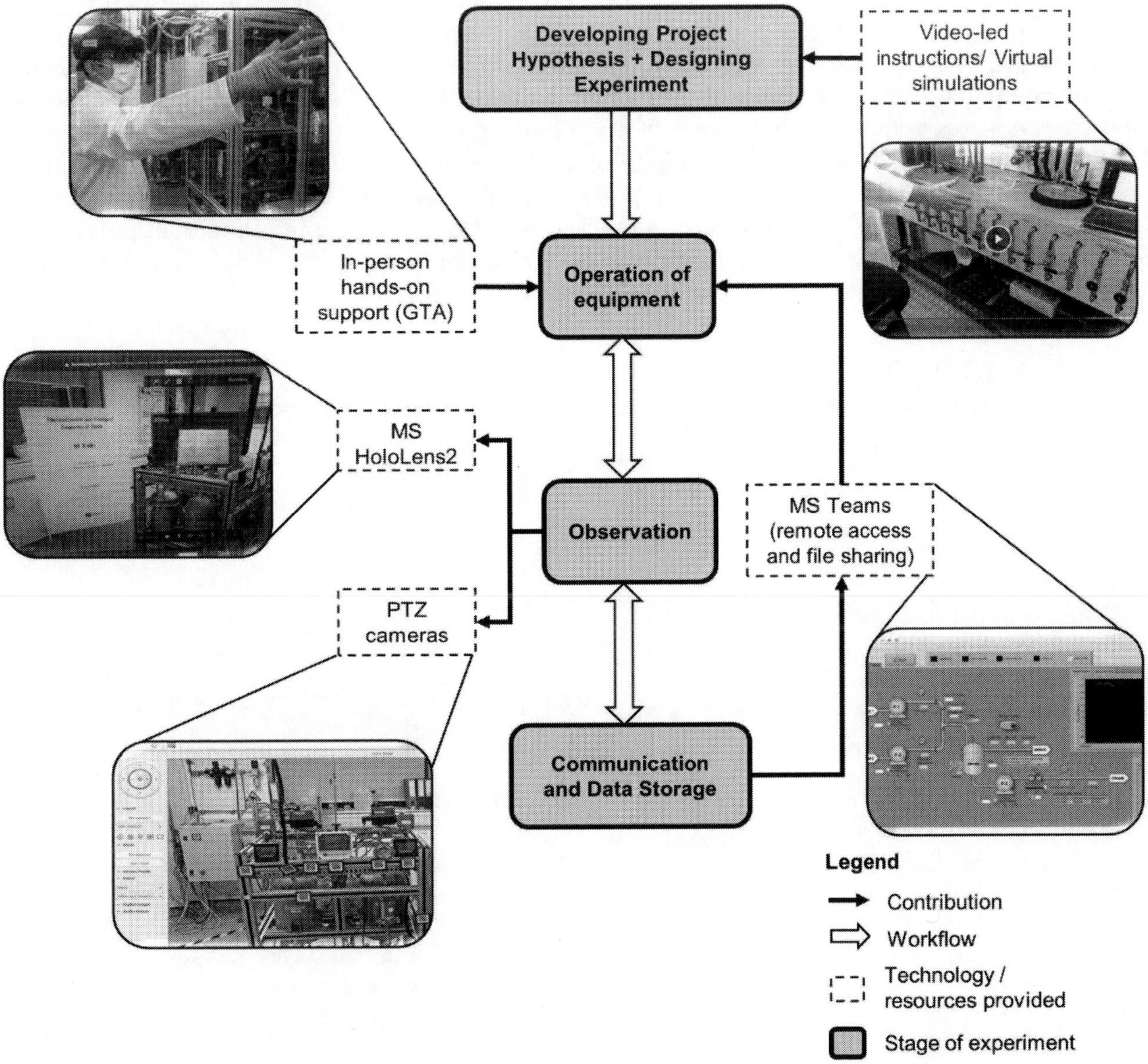

186g Providing Chemical Engineering Seniors an Industry-like Control Room Experience through a Remotely Operated Gas Permeation Experiment

Aravind Suresh, Jason Risolo, Emily Oldag

As part of the renovation of the chemical engineering laboratory at Manhattan College, the gas permeation experiment was redesigned for remote operation. The objective of the modification was to provide students a hands-on understanding of how process parameters are monitored and changed as needed from within a control room in a chemical engineering plant. State-of-the-art PRISM membranes were donated by Air Products to replace the existing membrane modules in the setup. Electronically operated on-off valves, needle valve, pressure regulator, pressure transducers, thermocouples, and oxygen sensors were plumbed into the experiment. A set of Programmable Logic Controllers (PLCs) and software were used to enable complete operation of the experiment using a desktop computer. Remote access software was then used to provide students taking the senior-level laboratory courses with scheduled access to the computer to carry out the experiment from a location inside the engineering building but outside the chemical engineering laboratory itself. In the first half of this presentation, the experimental setup will be explained, including details on how the experiment is carried out by the students. In the second half of the presentation, the presenter will provide a live demonstration of the experiment by running an operating condition in coordination with graduate assistants present in the physical laboratory located in New York.

186h Redefining the Scale of Fluid Flow Experiments

Fernando Mérida Figueróa, Carlos Rinaldi-Ramos, Spyros Svoronos

The sudden migration to online instruction was one of the repercussions of the global COVID-19 pandemic which brought unprecedent challenges for engineering education, especially for laboratory-based courses with hands-on learning outcomes. Unit Operations laboratories are one of these courses in Chemical Engineering undergraduate programs which often involves experimentation with pilot-scale flow systems comprised of pumps, valves, flow meters, and pipe networks, to study the principles of fluid flow in pipes. The approach behind fluid flow experiments relies on the direct operation of flow components by students, typically to characterize friction losses in pipes, fittings, and pumps. The lack of hands-on experience that would result from moving Unit Operations laboratories entirely online, without experimentation, was a big concern as it would have compromised some of the student learning outcomes thus jeopardizing the quality of instruction, and even program accreditation.

Motivated by the need to alleviate these challenges, small-scale experimental kits were conceptualized, designed, and utilized in online instruction of Unit Operations 1, the first of the three laboratory-based classes taken by undergraduate Chemical Engineering students at University of Florida during their junior and senior years. Experimental kits were designed to satisfy criteria such as modularity, cost-effectiveness, and safety, and to be easily shipped to each student in the class. Among these kits, the mini fluid flow (MFF) system was designed to have a 3D-printed fluidic device comprised of pipes of different diameters, contractions/expansions, bends, and built-in pressure taps. The fluidic device was connected to a submersible pump, receiving water from a plastic reservoir by means of plastic tubing containing flow reducers, and a small pinch valve to adjust the flow rate. Eleven "tube taps" enabled pressure drop measurements across pipes and fittings using a miniaturized differential pressure sensor that was subsequently connected to an Arduino microprocessor for real-time data monitoring and acquisition. Rigorous characterization tests were conducted not only to match operating ranges of both pump and pressure sensor but also to satisfy the desired scale of the system, among other design criteria. The MFF was designed to have a main inlet (receiving water from the pump) and two outlets discharging water back to the reservoir thus completing a closed loop. A plastic, graduated cylinder was used to measure flow rate using the so-called stopwatch and bucket method. The signal transferred from the differential pressure sensor to the Arduino microprocessor is read and saved in a personal computer via acquisition software. The overall goal of mini fluid flow experiments was the systematic investigation of friction losses using pressure drop measurements across multiple flow pathways. Experiments were designed such that students were able to conduct experiments individually at home, controlling experimental factors such as flow rate and flow configurations, and having the total pressure drop across the device as response variable.

The incorporation of mini fluid flow experiments in Unit Operations 1 when taught 100% online, demonstrated the ability to maintain high-quality, online instruction of fluid mechanics topics using small-scale, versatile experiments with opportunities for multiple experimental configurations and statistical analysis. Even though remotely assisted by instructors and lab assistants, students conducting experiments at home were able to fully assemble/disassemble the various MFF components which fostered troubleshooting abilities. A subsequent semester with "HyFlex" instruction (online + in-person) along with improvements in the design of the MFF system led to increased, positive acceptance by students, enhanced team dynamics, and overall improved understanding of topics learnt in lecture classes taken in previous semesters. A special highlight of experiments conducted in HyFlex classes was the flexibility to propose alternative experimental designs which fostered creativity and imagination among students.

Additional studies are currently underway to evaluate the potential of scaling-up and/or scaling-down fluid flow experiments by combining the MFF system with a pilot-scale pipe network, aiming to explore the ability of fluid flow systems to predict flow patterns in experiments regardless of the scale. In addition, further improvements to the MFF system will enable the characterization of fluid dynamics under fully turbulent flow thus strengthening the objectives of fluid flow experiments to be used as a single, small-scale systems or as hybrid (small- and pilot-scale) modules without the restrictions imposed by flow regime. Furthermore, the incorporation of the MFF system is suggested to enforce theoretical concepts of lecture-based courses and as an opportunity to showcase Engineering programs via K-12 experimental demonstrations, science fairs, and outreach initiatives.

302a Constructing a Syllabus with an Accessible Community in Mind

Jeffrey Halpern, Mariah Arral

Course syllabuses are students first introduction to the course content, and instructor teaching pedagogies. After several years of teaching Chemical Engineering courses and discussions with students about accommodations, we have found several small changes instructors can make to help create an accessible classroom. The first is making customized accessibility and accommodation statements. It is important to provide the specific language from your university/college on student resources, but often that language can be supplemented for your particular classroom community. In the past, we have added information from the American Test Anxiety Association and connected students to resources at our university in our syllabi. The second way instructors can incorporate accessibility is to create flexible extension policies. Instructors are often not aware of other pressing academic deadlines, or personal life events that students are balancing. As instructors, and researchers, we often ask for extensions on deadlines. Teaching students about asking for extension is a life skill and allows the goal of assignments to be about the students learning course material. Extensions are often a way that allow students to complete the required work, but not every student knows the culture and appropriateness in requesting an extension. Setting a specific framework transparent in the syllabus of when an extension is allowed, and how to request one, allow the student to finish their work in a timeframe that works best for them. Finally, the syllabus sets the tone of the classroom community. A syllabus set up for inclusion and positive learning clearly articulates that all are welcomed and sets clear boundaries on the learning community moving forward. Within this discussion, we will share examples of different syllabi and discuss the pros/cons of various inclusion statements.

J.M.H. was supported by a National Science Foundation EPSCoR award (#2119237).

M.L.A. was supported by an NSF Graduate Research Fellowship Program award (number DGE1745016)

302b Recalibrating Student Learning Goals Using Bloom's Taxonomy

Christy West

After a disappointing performance on a first test in reactor design, students were shown Bloom's Taxonomy and asked to indicate where they had been aiming in their preparation. Students thought that "understand" seemed the appropriate level of learning. (This may result from the frequent usage of "understand" in the course objectives on syllabi.) Upon being told that in a second-semester junior class their goal should be "analyze" or "evaluate," the students were able to recalibrate their efforts and focus on applying critical thinking abilities as they studied. This presentation will demonstrate how chemical engineering instructors can elaborate on Bloom's Taxonomy to communicate to their students specific examples of the level of learning expected in their courses.

302c Is Jango Fett's Backpack Thermodynamically Feasible? Bringing Students' Interest into Chemical Engineering Courses Using YouTube Problems

Matthew Liberatore

YouTube problems are visual, creative, student-written problems for any course. Developing an assignment for and grading of these unique problems that reverse engineer action in student-selected YouTube videos have resulted in a plethora of engaging and rigorous problems in courses including thermodynamics, material and energy balances, fluid mechanics, separations, and more. Our YouTube pedagogy has been in development for more than a decade, which has created an archive of hundreds of problems, some of which are freely available to faculty as part of the Material and Energy Balances zyBook. Additionally, our NSF-sponsored research studied problem-solving ability, measuring problem difficulty, and student attitudes, which will be briefly summarized here. Information on how to create YouTube problems, example rubrics, and our publications are available on our website as well as a YouTube channel title Reverse Engineering Engineering Education.

302d Improving Student Writing: Drafts and Reflections on Lab Reports

Katie Cadwell

A common lab instructor situation: spend hours upon hours grading incomplete lab reports whose authors have seemingly ignored all provided requirements and detailed grading rubrics. Leave loads of helpful comments, only to have students acknowledge that they only looked at the numerical score - sub-scores if you're lucky - and then go on to make the exact same errors in their next report. I've added two minor assignments to my chemical engineering lab course to address this frustrating experience.

First, a week in advance of each report date students submit the roughest of drafts that is a narrative outline of the full report, with sentences like "...as the concentration increases the viscosity *[increased/decreased/stayed constant]*, which *[agrees/disagrees]* with the theory found in *[need source - maybe fluids textbook?]*." Then via an in-class activity, other students review the draft for completeness (has the author addressed each required part of the analysis, even if the details are still TBD?) and for comprehensibility (grammar, sentence structure, organization). In this way writers and their peers identify in advance which parts of the report remain to be analyzed, written or edited. After reports are graded, I assign a reflection assignment on the marked report. Students have to read through my comments to identify what I complimented them on, what I recommend they do different, and then report in a memo on what they will do the same or different in the next such report.

The draft, activity participation, and reflection are all graded on the basis of complete/incomplete, and so does not lead to significantly more time spent grading. Since implementing these assignments, I'm receiving fewer woefully incomplete report submissions, student memo content indicates that students are actually reading my copious feedback, and at least sometimes they use that feedback to improve their subsequent reports. Additionally, the draft prevents students from procrastinating beginning the writing portion of their reports until the day it's due, and the reflection memo has helped me to stop procrastinating grading lab reports since they must be returned to the students in advance of the memo deadline.

302e Art Conservation and Chemical Engineering Product Design

Rebecca Harmon

Chemical engineering reaches into many different fields, even art conservation and the broader field of cultural heritage science. A two-day synchronous module was developed for an upper-level product design course. An art conservation scientist was interviewed and took questions from the class on techniques and problem-solving experiences with art objects. A case study followed on applying the design process to a current art conservation problem of stabilizing flaking paint given several client and chemical constraints. This module offered a look at a non-traditional application of chemical engineering fundamentals with a mix of art and engineering, which may catch the interest of students to expand their perspective on technical career paths.

302f Let's Talk about Process Safety: Utilizing an Online Forum-Based Approach to Discuss Process Safety Incidents

Richard Cimino

Traditionally, the process safety course taught at our University utilized substantial in-class time to watch process safety incident videos published by the Chemical Safety Board (CSB.gov). These videos were followed by extensive, interactive in-class discussions on the particular process safety topics and recommendations covered in the videos. However, the global COVID-19 pandemic made this mode of instructional discussion untenable. In particular, the ability for students to participate in in-class discussions and to express themselves about process safety topics was severely hindered by the virtual nature of instruction. To work around this issue, a commonly used technique for asynchronous-online instruction was implemented to allow the students to interact with this material, albeit on their own time. Utilizing the Canvas learning management system's built-in Discussions tool, students now watch one topical video per week asynchronously and participate in online forum-style discussions about the material. Students are required to respond to a set of prompting questions (supplied by the instructor), and in addition, each student must respond to at least one other student's initial post. This latter requirement ensures that the students are interacting and reading each other's posts. The weekly discussions culminate with a capstone discussion, which requires students to research and analyze historical Ammonium Nitrate accidents and compare them to the Beirut harbor explosion in 2020. These discussions are used to evaluate the students' ability to recognize the social and ethical dimensions of process safety, including the impact of process safety incidents on societal and global scales, pursuant to ABET Learning Outcome 4B.

384a A State-of-the-Art Pilot-Scale Distillation Column for the Unit Operations Laboratory at the University of Kansas

Felipe Anaya, David M. Griffin, Mark B. Shiflett

A new pilot-scale distillation column has been commissioned for the Unit Operation Laboratories at the University of Kansas (KU) and will begin official laboratory course operations in the Spring of 2023. The column is 6.3 in. diameter, nearly 20 ft. tall, made from stainless steel, and with a total of 20 conventional sieve trays. The addition of this equipment is the most recent step in a concerted effort to update, modernize, and improve all experiments with the goal of enhancing our students' learning experience. The Unit Operations Laboratory course at the University of Kansas is designed as a two-part course taken during the Fall and Spring semesters of the senior year. The courses are titled Chemical Engineering Laboratory I and II for the Spring and Fall semesters, respectively. In the first course of the series, students cover experiments on the fundamentals of Chemical Engineering learned during their Junior year (Kinetics, Fluids, Thermodynamics). In the second laboratory course, students cover Unit Operations experiments such as absorption and distillation and other aspects of the senior curriculum such as process dynamics and control.

The new distillation column is part of the second laboratory in the course series and is replacing a previous bench scale glass distillation column used for many years. With the projected upgrades, the UOL facilities will have pilot and bench-scale equipment located in the laboratory's high and low bay area, respectively. So far, 4 rack-mounted bench scale experiments have been commissioned and constructed in the low-bay area. These racks are built on wheels, so they are mobile. This feature allows rolling the experiments into the classroom where the students are taking the associated theory class to introduce the new experiment before beginning the lab their senior year. The distillation column is one of the pilot-scale experiments located in the high-bay area. This high bay has a dedicated control room from which the distillation column will be operated. The column is highly automated with variable frequency drive pumps and automatic control valves. The distillation column will be operated and controlled with a Programmable Logic Controller (PLC) linked to a desktop computer running LabVIEW. The control room is furnished with industry-like console workstations and 27 inch dual monitor computers. Additionally, tablets are wirelessly connected to the PLC to operate the distillation column while doing walkthroughs in the high-bay area. Additional features include cameras installed on 5 different sight glasses to visually monitor the activity of the trays at different points in the column from the computer in the control room. Very few operational steps will require being physically present in the high bay so that the column can be operated remotely. In this presentation, details of the distillation column design, construction, and operation will be presented along with planned strategies to optimize the utilization of column features to enhance student learning.

384b Design Your Own Experiment: An Alternative Experiment for Seniors in Process Engineering Laboratory

Alex Bertuccio

Process Engineering Laboratory (PEL, also known as Unit Operations or Senior Lab) is taught as a two-semester course to provide students with experience working with different processes and understanding fundamental phenomena. Experiments are open-ended to encourage students to explore numerous variables. At the conclusion of each experiment students submit a team lab report and give an individual presentation. Students repeat this experimental cycle several times throughout the year to develop their analytical and communication skills. However, by the second half of the spring semester, seniors often have reduced motivation and enthusiasm for conducting experiments. This year, for the final experiment, students were challenged to design and test their own engineering experiment for PEL. Experiments needed to be safe, cost-effective, open-ended (with a multiple variables that could be tested), and quantifiable. This talk will explain the structure of this experiment and show several experiments designed by students.

Design Projects in the Unit Operations Laboratory

Stephanie D. Loveland[1] and Kristin L. Stoner[2]
[1]Department of Chemical and Biological Engineering
[2]Department of English
Iowa State University, Ames, Iowa

Introduction:

Proposals are a staple in advanced communication courses due to their commonality in the majority of professional realms. Due to its relevance, an informal proposal is part of the standard curriculum for both business communication and technical communication courses at Iowa State University, courses required in the majority of campus programs.

In advanced communication, students are asked to demonstrate the understanding and practice of the rhetorical concepts and processes to successfully communicate individually and collaboratively in the four modes (written, oral, visual, and electronic) across a range of disciplines. More specifically, technical communication emphasizes strategies for analyzing expert and lay audiences and adapting information to those audiences. It includes documentation, visualizing data, and presenting information orally. In the English link for chemical engineering, students apply these goals and practices to lab reports, documenting experiments carried out in their chemical engineering lab. The majority of the course focuses on audience analysis for successful expository communication in all four modes.

In the unit operations laboratory, students are asked to effectively collect and analyze data, develop appropriate experimental designs for data collection, and consider the safety and environmental impacts of that experimental work. The laboratory course is also well-suited to practice and improve the variety of communication skills, including written and oral.

In this paper, a design project in the Unit Operations laboratory is described. This project is an informal proposal presented to the instruction staff for both the laboratory and the technical writing courses, who jointly assess the presentation and related documents.

.

ChE 325 and English 314: the Technical Writing Learning Community

The first Unit Operations laboratory course at Iowa State University has been linked with a technical writing course since 2012. The means of creating the link and the learning community have been described in presentations by Loveland and McGough [1] and Loveland and Satterwhite [2] [3]. Through this learning community, students have the opportunity to learn several different modes of communication that are relevant to the field of chemical engineering, and to have their work assessed by faculty with differing expectations for that work. Most of the assignments given in the laboratory involve some form of report to convey the results of experimental tests. In this learning community, those reports are submitted to the instructors for both classes.

The assignments in the linked course are scaffolded according to the modes of communication and the expertise needed to successfully execute those modes to intended audiences. The scaffolding also mirrors the lab where students are prompted to consider the variety of situations in which their research is communicated, including conference posters and oral presentations. The proposal presentation assignment is the culmination of those skills acquired in both courses with the added task of persuasive rhetoric applied to an industry specific topic.

The Proposal Project

The Final Proposal Presentation Assignment asks students to collaboratively propose an experiment be added to the chemical engineering lab curriculum. Their focus audience includes their peers, their chemical engineering instructor, their English instructor, and Chemical Engineering Department administration. This variety of technical and lay audiences emphasizes the scaffolding of both courses' content. The product produced is a slideshow and fifteen-minute oral presentation. The final product should demonstrate an understanding of lab practices, chemical engineering course outcomes, chemical engineering industry standards, research techniques, audience analysis, persuasive rhetoric, oral communication practices, and visual communication practices.

Students are first introduced to the assignment in their lab when asked to consider a variety of possible experiments. With their lab partners, they select their top three choices for an experiment for which they can argue from a list of possible experiments provided by the laboratory instructor. Prior to making that selection, they are introduced to the assignment specifics in their English course. They first discuss the different types of proposals, internal, external, solicited, and unsolicited. They are then asked to identify which type of proposal they are being asked to develop for this assignment based on the assignment information made available to them via a detailed assignment sheet and rubric. Students then begin to analyze audience expectations for the internal, solicited proposal for this assignment. They then submit their requests for experiment topics to their chemical engineering instructor. Once students have been assigned their proposal topic, they continue to prepare for the presentations in both courses.

In the lab they focus on accuracy of information, theories tested, and required materials. They must design the proposed experiment to meet several constraints, including cost and space limitations, and state specific learning goals for the new experiments, explaining how those goals meet the laboratory course curriculum requirements. limitations. Students consider inherently safer design strategies as they design the equipment and materials to be used and explain how any waste materials will be handled, as well as what environmental or safety concerns might exist with the materials being used. They will also address any physical hazards such as pressure, temperature, and rotating equipment. They are reminded that since this course is the first time many students will work with real equipment, the hazards must be minimized in order to prevent the inexperience of the users from contributing to an accident. Finally, they will demonstrate the expected results of the experiment, and how they expect future students to analyze the data form their proposed experiment.

In their English course students use all of this information to develop a strong argument, considering rhetoric, appeals, and strong presentation and visual communication practices. In the end, they appeal to the instructors to incorporate their design as a new experiment in the laboratory.

Proposal Assessment

During the presentations, both instructors assess student performance and product by utilizing the same rubric. The chemical engineering instructor focuses on those criteria that assess accuracy of information and depth of content, while the English instructor assesses criteria specific to audience analysis and communication techniques. For students, this dual grading method seals those seams between the two courses, already tightly knit by previous course content. It demonstrates that in the professional realm they will be evaluated beyond specific field expertise; to be successful that expertise must be communicated in a variety of situations to a variety of audiences.

Although all assignments in these linked courses are unique in their strong demonstration of the critical connection between engineering and communication, the Final Proposal Presentation is the apex of those courses' collaboration and scaffolding.

Summary/Conclusions

The technical writing learning community that links the first of two unit operations laboratory courses with a technical writing course has become a vital part of the chemical engineering curriculum at Iowa State University. Students who participate receive training in the variety of communications required for them as chemical engineers, and the Final Proposal Presentation Assignment is a culmination of the concepts practiced throughout the semester.

References

[1] S. D. Loveland and S. D. McGough, "A Learning Community: Teaching Technical Writing in the Chemical Engineering Unit Operations Laboratory," in *Proceedings of the ASEE North Midwest Section Conference*, Iowa City, IA, 2014.

[2] S. D. Loveland and M. Satterwhite, "Teaching Technical Communication to Chemical Engineering Students Through the Use of a Learning Community and Linked Courses," in *AIChE Annual Meeting Conference Proceedings*, Pittsburgh, PA, 2018.

[3] S. D. Loveland and M. Satterwhite, "Communicating with Both Engineers and Non-Engineers: Incorporating Multi-Disciplinary Communication Strategies in the Unit Operations Laboratory," in *AIChE Annual Meeting Conference Proceedings*, [virtual conference], 2020.

[4] ABET, "Criteria for Accrediting Engineering Programs, 2022-2023," 2021. [Online]. Available: https://www.abet.org/accreditation/accreditation-criteria/criteria-for-accrediting-engineering-programs-2022-2023/. [Accessed 7 November 22].

384d Enhance Safety Knowledge and Mindset through the ChE Undergraduate Laboratory Courses

Chijuan Hu

Transport lab and Unit Operations lab, two ChE undergraduate core courses, are taught in a laboratory environment which involves various types of hazards (e.g. chemicals, biologicals, electricity, and mechanical). As noted in the ABET expected learning outcomes, students must develop the abilities of producing solutions and recognizing the responsibilities to conduct work with the considerations of safety. These laboratory courses serve as a unique forum to practice and assess the achievement of this goal. In this presentation, I will discuss how safety concepts are taught, practiced, and assessed in the ChE laboratory courses at Texas Tech.

The following teaching pedagogies and media have been applied to enhance students' knowledge and awareness of safety: 1). Provide dedicated lectures to discuss safety protocols and safety precaution needed; 2). Give a virtual tour of the lab and locate the emergency equipment prior to entering the lab space; 3). Create videos to explain the equipment setup and identify areas for precaution; 4). Share lessons learned from past incidents and facilitate discussions on preventative measures (e.g. videos from Chemical Safety Board); 5). Introduce procedures on how to respond to emergency such as fire, severe weather, and medical emergency; and 6). Distribute safety materials like wallet-sized safety card of useful links and emergency contacts. Various assessment methods are applied as follows: 1). Complete the university required online safety training; 2). Pass an in-class safety exam based on the safety lecture; 3). Give pre-lab presentation to identify the hazards involved and provide risk mitigation plans; and 4). Evaluate safety practices in the lab performance as stated in the grading rubrics.

Some new ideas of developing ways to enhance the safety aspect in the ChE education are discussed: 1). Hands-on situational training can be more effective and long-lasting in memory; 2). Require a short group meeting to have a safety briefing and filling out job safety analysis form to simulate this common practice in industry; 3). Hold a panel discussion from alumni to talk about safety culture and practices in their current organizations; and 4). Create a project related to the experiments in a real-life scenario where students are expected to conduct research and apply the safety knowledge to prevent a foreseeable incident.

384e Continued Improvement of the Chemical Engineering Undergraduate Laboratories at the University of Texas at Austin

Carlos Landaverde-Alvarado

The McKetta Department of Chemical Engineering at the University of Texas at Austin started an ongoing revitalization of the two required chemical engineering undergraduate laboratories during the 2020-21 academic year. The main purpose of this ongoing revitalization has been to update and improve the content, teaching effectiveness, and general learning experience offered by the undergraduate laboratory sequence at UT Austin (Junior and Senior year). Improvements to the laboratories are based on the recommendations of a faculty committee that assessed the courses, feedback from the faculty and staff associated with the laboratories, as well as input received from industry partners and the advisory board of the department. As a consequence of this process, we have implemented several changes with aims to continuously improve the practical learning experience of chemical engineering students as they achieve the overall objectives of the laboratories. Furthermore, and as part of this process, the courses have greatly benefited from our collaborations with and the input we receive from industry, and the participation of staff, faculty, and graduate students in the development of the new learning experience offered to students.

The objective of our current revamping efforts has been to greatly enhance the educational laboratory experience for every single chemical engineer who graduates from the University. By the implemented improvements, amongst other positive outcomes, students have been able to explore practically more areas of chemical engineering and have now more tools available that enhance their hiring profile for different industries. Moreover, the laboratory courses have become one of the main windows of our department to create collaborations between our students and industry leaders, industry partners and department alumni. With this in mind, the ultimate goal of this revitalization process is to continue making the labs a prize component in the department of Chemical Engineering that ties together all components of the chemical engineering curriculum, reinforces chemical engineering concepts, and prepares students for their careers post-graduation.

The updates implemented to this day have resulted in the addition of several new experiment stations, the update of all existing experiment stations and laboratory manuals, the revision of the course curriculum, and the increase of industry collaborations that enhance the learning experience of our students. This ongoing process intents to continuously evaluate the content of the courses and ensure they serve effectively the current needs of chemical engineering students, chemical engineering research, and the chemical engineering industry along with other stakeholders of our program. The laboratories have also benefited from an amplified emphasis on process and laboratory safety as a result of these updates.

The overarching goal of this presentation will be to provide an overview of the laboratory sequence at UT Austin and to present a brief detail of the recent improvements to the laboratories. In addition, we will describe two of the recently added experiment stations that focus on an enhanced design experience in the class projects (fluidized bed and separations), and a description of our updated approach to safety, in collaboration with industry partners of the department, that pursues the creation of a renovated safety culture within our program.

384f A Novel Approach for Laboratory Experiments in Process Dynamics and Control

Felipe Anaya, David M. Griffin, Mark B. Shiflett

A novel, state-of-the-art Process Dynamics and Controls experiment has been commissioned, constructed, and implemented for the Unit Operations Laboratory (UOL) at the University of Kansas (KU). For the last several years, the Chemical andamp; Petroleum Engineering Department at KU has mobilized resources to improve, update, and modernize the Unit Operations Laboratory experiments with the goal of enhancing the student's learning experiences. A novel approach to UOL instruction was implemented and several new experiments have been commissioned. This new approach has received positive feedback from students and resulted in improved learning outcomes. Unit Operations Laboratory courses at KU are designed as a series of two courses taken in the Fall and Spring semester of senior year: Chemical Engineering Laboratory I and II, respectively. In the first of the series, students cover 3 experiments on the fundamentals of Chemical Engineering (Kinetics, Fluids, Thermodynamics). For this course, 3 new and updated experiments were added and results on their implementation in the course structure were published in the Journal of Chemical Engineering Education. In the second laboratory course, students carry out three experiments on distillation, absorption, and process dynamics and control.

The process dynamics and control experiment discussed in this presentation is the most recent addition to our UOL facilities and it is meant for students in the second laboratory of the series. The system is assembled in a mobile rack so that it can be wheeled into the classroom where process control is taught prior to taking this laboratory course. With this apparatus, students can design and carry out experiments on liquid level control, pH control, or temperature control. The equipment consists of a set of two analogous "reservoir" glass tanks labeled "hot" and "cold" tanks (or "basic" and "acidic" tanks for pH experiments). Each of the reservoir tanks has an inlet water and an outlet water flow. The inlet water line to the hot tank goes through and electric heater before entering the tank. The outlet flow from both reservoir tanks are mixed and sent into a third "mixing" glass tank. Finally, the outlet flow of the mixing tanks is sent into a final collection tank. Additionally, mildly acidic or basic solutions can be continuously added into the reservoir tanks to make each tank either acidic or basic. Each of the three tanks is equipped with agitators as well as RTDs, pH probes, and level sensors. All flows described are controlled with either a pneumatic control valve, proportionating solenoid control valve, or variable frequency drive pump. Once switched on, the equipment is completely operated from a computer running LabVIEW and could be operated remotely. In this presentation, we will share details of the equipment design and associated educational advantages for the instruction of a laboratory in process dynamics and control. Positive feedback was received by the first class of students to operate the equipment in the Spring of 2022. The flexibility of the design allows for different options and teaching strategies. We will discuss some of the many possibilities to explore with future classes that will apply and deepen their knowledge of process dynamics and control with this experiment.

384g Green Cubesat Propulsion: Using Hands-on Experimentation to Develop Knowledge and Interest in STEM and NASA Career Opportunities.

Greg Ogden

Finding sustainable methods of spacecraft propulsion is an essential element in the continuation of space exploration. CubeSat propulsion systems are severely space constrained posing a challenge for researchers to find sustainable yet practical propellants. Historically, CubeSats were used as low-cost tools for science and research missions. They have also been touted as an exciting hands-on way to involve high-school students in space missions. As CubeSat use has exploded during the past decade, micro propulsion systems became necessary to facilitate guidance navigation and attitude control (GNAC), especially past low earth orbit (LEO). These GNAC systems need a compact propulsion systems that use safers and environmentally friendly propellants.

This presentation discusses CubeSat propulsion research conducted at the University of Arizona over the past three years as part of a NASA MUREP project aimed at educating students through hands-on experiences and meet NASA education goals by strengthening the future workforce for NASA and the nation. The challenges of providing hands-on experimental opportunities during COVID mandated school closure as well as the transition back to socially distanced on-campus activities will be discussed.

Graduate and Undergraduate students explored several CubeSat propulsion technologies using water as the propellant including solar thermal propulsion and resistojet thrusters up to 0.5N. Other groups investigated electrolyzer processes for on-board production of hydrogen and oxygen propellants for engines up to 2N. In this configuration, water is considered as a propulsion storage medium. The presentation will discuss developing system criteria, experimental design and construction and experimental results. Use of low-cost, commercially available components such as educational PEM electrolyzers was emphasized in line with the CubeSats initial mission as a low-cost easy access to space technology. Common pitfalls encountered by the teams will be presented and how each team overcame roadblocks such as ordering delays, equipment malfunctions and system scaling.

303a The People Who Get to Do the Science Determine What Science Gets Done: Elucidating the Impact of Red Blood Cell Rigidity in Sickle Cell Disease

Omolola Eniola-Adefeso

Racial Diversity remains lacking in the Engineering and Science faculty ranks. This lack of representation of Black, Hispanic/Latino/a, and Indigenous faculty researchers creates a significant burden for the US scientific enterprise both in our ability to generate science and technology that works for all Americans and our global competitiveness, which I will motivate with my academic journey and research at the University of Michigan.

303b Entrepreneurial Bioengineering Research and Design Via Innovative Collaboration

Mark Byrne

The inclusion of collaborative research and design opportunities within undergraduate and graduate curricula leads to enhanced learning in a unique environment. Students in multi-tiered teams exercise hands-on, open-ended problem solving to investigate contemporary bioengineering problems with external clinical and industrial partners. Common research and design elements can provide a needed curricular and assessment framework while at the same time creating experiences unique to each team. Students can substantially benefit by navigating external partner relationships and obtaining major deliverables.

303c The Role of Bioengineering in an Evolving Chemical Engineering Curriculum

Efrosini Kokkoli

In the 1970s and 80s, chemical engineering strongly influenced the development of a bioengineering curriculum on kinetics, transport phenomena, and thermodynamics. In the 1990s, bioengineering had matured into a distinct field with its own curriculum and teaching methodologies. At the same time, the chemical engineering started focusing more on molecular processes. Considering the ongoing evolution of our profession in the 1990s and 2000s, it became apparent that we had to educate our students on the molecular sciences, nanotechnology, and bioengineering. AIChE established the Society for Biological Engineering as a technological community, promoting the integration of chemical engineering with biology. Textbooks of our core classes included concepts and examples from both chemical and biochemical/biomolecular engineering. This presentation will discuss examples of this evolution and how bioengineering influenced the chemical engineering curriculum at Johns Hopkins University.

303d Modern Bioengineering and Its Home in Chemical Engineering

Hal Alper

It was once said, "the more things change, the more they stay the same." While the tools, applications, and problems in bioengineering certainly embrace a changing ability to understand and engineer biology, they also draw upon a solid foundation of chemical engineering principles and paradigms. One needs to look no further than then the person of honor for this session, Prof. Nicholas Peppas, to recognize the importance of understanding past scholarly approaches in order to trailblaze new paths forward. This talk will highlight the past, current, and future home for modern bioengineering within the discipline of chemical engineering with an emphasis on teaching, mentorship, and research.

303e Inclusive and Equitable Teaching Methods in Chemical Engineering

Shelly Peyton

In this special session honoring the Dr. Peppas, I will discuss several strategies to make classrooms and research labs more inclusive, equitable spaces for students from all racial and demographic backgrounds to thrive. Advancements in Chemical Engineering depend on our ability to recruit and retain the best students that have broad ranging perspectives they can bring to challenging problems in human health, energy, and the environment. It is our responsibility as educators to meet these needs of these students so that they can contribute to our dynamically changing field.

AUTHOR INDEX